The Theory of Numbers

The Theory of Numbers

An Introduction

Anthony A. Gioia
Western Michigan University

MARKHAM PUBLISHING COMPANY
CHICAGO

MARKHAM MATHEMATICS SERIES
William J. LeVeque, Editor

Lectures in Advanced Mathematics

Copyright © 1970 by Markham Publishing Company
All Rights Reserved
Printed in the U.S.A.
Library of Congress Catalog Card Number: 74–91021
Standard Book Number: 8410–1013–7

To Pat, Kathy, Mickey

PREFACE

THIS TEXT was developed for a first course in number theory at Western Michigan University, where students may take this course as early as their junior undergraduate year or as late as their first graduate year. With the diversity in the levels of the students' previous training in mind, I selected the topics and methods of presentation for this text. Every chapter is written under the assumption that the student's background consists only of calculus (including the brief introductions to analytic geometry and linear algebra). The instructor can achieve various depths in the course, in order to compensate for the differing backgrounds, by requiring individual students to complete exercises of a degree of difficulty proportionate to the student's proficiency.

Because of the assumption of only a post-calculus standard of maturity of the student, any of the chapters may be presented in any trimester regardless of variations in preparedness of the audience from one trimester to another. (To aid the instructor in the selection of chapters to be presented, a table of logical dependence of the sections is given at the end of this Preface.) Moreover, regardless of the extent of the student's prior training, his sophistication increases only slightly in one trimester; accordingly, the rate of presentation of material in this text remains almost constant throughout.

Arithmetic functions are introduced early (Chapter 2) and are used often in the development (Chapter 4; Chapter 8; Section 14 of Chapter 3; and Section 30 of Chapter 5). The emphasis is on the algebraic structure of the set of functions under Dirichlet convolution; although this has been common in the literature, it has received inadequate stress in published texts. The algebraic viewpoint has the obvious advantage of leading to the development of a calculus which eliminates the exigency of proofs replete with mysterious combinatorial and unpleasant computational techniques.

In Chapter 8 the Erdös-Selberg proof of the Prime Number Theorem is given. In part, my reason for choosing to give the elementary proof rather than a nonelementary proof of the Prime Number Theorem is that the elementary proof comes as an easy (though not short) application of methods already used in Chapter 4 to derive the standard results on orders of magnitude, namely, the Euler-McLaurin sum formula, summation by parts, and changing the order of summations. And in part my reason is that the under-graduate rarely has the requisite background in function theory to follow a nonelementary proof.

In addition to the introductions to the theory of arithmetic functions and analytic number theory cited above, the text also gives introductions to algebraic and geometric number theory. The former is furnished by the study of the Gaussian integers (to enumerate the ways of representing a positive integer as a sum of two squares), and the Jacobian integers (to prove the insolvability of the Fermat cubic). Because of the emphasis on the similarity of methods used for the rational, Gaussian, and Jacobian integers, the student can appreciate that these are but three special cases of a general theory. The introduction to geometric number theory is supplied by the use of geometric methods in the proof of the Quadratic Reciprocity Law (the concept of a lattice point is introduced in Section 17); in the proofs of certain asymptotic formulas for summatory functions (in Sections 19, 21, and 22); and, of course, throughout Chapter 9.

I wish to thank Professor John Petro and Dr. A. M. Vaidya for reading the manuscript, pointing out an embarrassing number of errors, and helping to supplement the original lists of exercises; Professor William J. LeVeque for his valuable suggestions about the final chapters; and Mrs. Judith Warriner for typing the manuscript.

A.A.G.

September, 1969
Kalamazoo, Michigan

LOGICAL DEPENDENCE OF SECTIONS

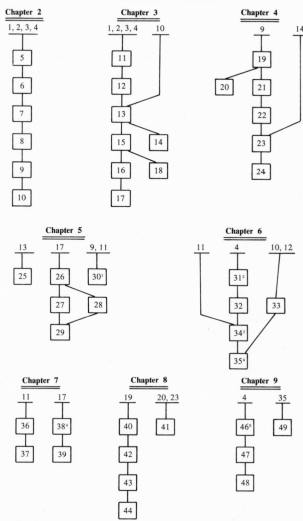

[1] Need the statement of Theorem 27.1.
[2] Need Section 11 for Exercises.
[3] Need Section 17 for Exercises.
[4] Need Lemma 26.2, which can be proved after Section 10.
[5] Need Section 17 for an example.

CONTENTS

Chapter 1

FUNDAMENTAL CONCEPTS

1. Divisibility

Let $\mathscr{Z}$ denote the set of integers, that is,

$$\mathscr{Z} = \{\ldots, -3, -2, -1, 0, 1, 2, 3, \ldots\},$$

and $\mathscr{Z}^+$ denote the set of positive integers, $\mathscr{Z}^+ = \{1, 2, 3, \ldots\}$. The following definition of divisibility introduces a relation on $\mathscr{Z}$ which is basic to most of the material presented later.

Definition. *Suppose $a, b \in \mathscr{Z}$. We say a divides b (or b is a multiple of a), and write $a|b$, if and only if there exists $c \in \mathscr{Z}$ such that $ac = b$. If a does not divide b, we write $a \nmid b$.*

Notice that if $b \neq 0$, then $a|b$ implies that $a \neq 0$; equivalently, if $a = 0$ and $a|b$, then $b = 0$. However, if $b = 0$, then $a|b$ for *every* $a \in \mathscr{Z}$. Some further consequences of the definition are given in the theorems below.

Theorem 1.1.

(a) *For every $a \in \mathscr{Z}$, $a|a$.*
(b) *If $a|b$ and $b|a$, then $a = \pm b$.*
(c) *If $a|b$ and $b|c$, then $a|c$.*
(d) *If $a|b$ and $a|c$, then $a|(bx + cy)$ for all $x, y \in \mathscr{Z}$.*

Proof. The proofs of (a), (b), and (c) are left as exercises. To prove (d) we observe that $a|b$ and $a|c$ implies the existence of $d, e \in \mathscr{Z}$ such that $ad = b$ and $ae = c$. But then $a(dx + ey) = bx + cy$ for all $x, y \in \mathscr{Z}$, and since $dx + ey \in \mathscr{Z}$, we have $a|(bx + cy)$, as required.

Theorem 1.2. *If $b \neq 0$ and $a|b$, then $|a| \leq |b|$.*

Proof. For some $c \in \mathscr{Z}$, we have $ac = b$. Then $|a||c| = |b| \neq 0$. Thus, $|c| \neq 0$, so $|c| \geq 1$. Therefore, $|a| \leq |a||c| = |b|$.

1

Theorem 1.3. *If $a \in \mathscr{Z}$ and $b \in \mathscr{Z}^+$, then there exist unique $q, r \in \mathscr{Z}$ such that $a = bq + r$, $0 \le r < b$.*

Proof. Consider the set

$$\mathscr{S} = \{a + bx : x \in \mathscr{Z}\}.$$

If $a \ge 0$, then $a + b \ge 0$ and $a + b \in \mathscr{S}$; if $a < 0$, then $a + b(-a) = a(1 - b) \ge 0$ and is an element of $\mathscr{S}$. In either case $\mathscr{S}$ contains non-negative integers, and therefore contains a smallest non-negative integer, say r. Suppose that r occurs in $\mathscr{S}$ with $x = -q$, so we have

$$0 \le r = a - bq.$$

If $r \ge b$, then $a - bq \ge b$, from which it follows that

$$0 \le a + b(-q - 1) < a - bq.$$

But this is impossible since $a - bq$ is the smallest non-negative integer of the form $a + bx$. Hence, $b < r$ and there are integers q and r as stated in the theorem.

To prove uniqueness, we assume there are integers q, q_1, r, r_1, such that

$$a = bq + r, \qquad 0 \le r < b$$

and

$$a = bq_1 + r_1, \qquad 0 \le r_1 < b.$$

Now assume that $q < q_1$. Then

$$0 \le r_1 = a - bq_1 \le a - b(q + 1) = r - b < 0,$$

an obvious contradiction. Similarly, $q_1 \nless q$. Hence $q = q_1$, and

$$r_1 = a - bq_1 = a - bq = r.$$

Corollary 1.3. (*The division algorithm.*) *If $a \in \mathscr{Z}$ and $b \ne 0$, $b \in \mathscr{Z}$, then there exist unique $q, r \in \mathscr{Z}$ such that*

$$a = bq + r, \qquad 0 \le r < |b|.$$

Proof. If $b > 0$, the corollary is the same as Theorem 1.3. If $b < 0$, then a and $|b|$ satisfy the hypothesis of Theorem 1.3, so there exist $q', r \in \mathscr{Z}$ such that $a = |b|q' + r$, $0 \le r < |b|$. But with $b < 0$, $|b| = -b$; take $q = -q'$ and the corollary is proved.

EXERCISES
1-1. Prove Theorem 1.1 (a), (b), and (c).
1-2. Show that Theorem 1.2 is not necessarily true if $b = 0$.

1-3. Suppose $\mathscr{S}$ is a set and $\mathscr{R}$ is a relation on $\mathscr{S}$. If r, $s \in \mathscr{S}$, we write $r\mathscr{R}s$ to mean that r is in the relation $\mathscr{R}$ to s. The relation $\mathscr{R}$ is called a *partial order* on $\mathscr{S}$ provided (a) $s\mathscr{R}s$ for all $s \in \mathscr{S}$, (b) if $r\mathscr{R}s$ and $s\mathscr{R}r$, then $r = s$, and (c) if $r\mathscr{R}s$ and $s\mathscr{R}t$, then $r\mathscr{R}t$. Prove that "$|$" is a partial order on $\mathscr{Z}^+$.

1-4.* Prove by mathematical induction that if $a|b_1, a|b_2, \ldots, a|b_n$, then $a|(b_1x_1 + \cdots + b_nx_n)$ for every set $\{x_1, x_2, \ldots, x_n\} \subset \mathscr{Z}$. Thus prove that if $a|b_i$ ($i = 1, \ldots, n$) and if c can be written as a linear combination of the b_i (that is, there are integers $x_1, \ldots, x_n$ such that $c = b_1x_1 + \cdots + b_nx_n$), then $a|c$.

1-5.* Prove
  (a) If $a|b$, then $a|bc$ for every $c \in \mathscr{Z}$.
  (b) If $a|b$, then $ac|bc$ for every $c \in \mathscr{Z}$.
  (c) If $a|b$ and $c|d$, then $ac|bd$.

1-6.* If $a \neq 0$, show that the set $\{d : d|a\}$ is a finite set.

1-7.* Suppose $n \in \mathscr{Z}^+$ is fixed, and let

$$\mathscr{S} = \{d : d \in \mathscr{Z}^+, d|n\}.$$

Prove
  (a) $d \in \mathscr{S}$ if and only if $n/d \in \mathscr{S}$.
  (b) If the elements of $\mathscr{S}$ are ordered by increasing magnitude, $1 = d_1 < d_2 < \cdots < d_t = n$, then the corresponding elements n/d_i ($i = 1, \ldots, t$) are ordered by decreasing magnitude.

1-8. Suppose $a, b \in \mathscr{Z}^+$ and $ab = c$. Prove that either $a \leq \sqrt{c}$ or $b \leq \sqrt{c}$.

1-9. An integer n is even if $2|n$, odd if $2\nmid n$. Prove that the sum and difference
  (a) of two even integers is even,
  (b) of two odd integers is even,
  (c) of an even and an odd integer is odd.

1-10. If n is an odd integer different from ± 1, prove that n cannot divide both of two consecutive even integers, and that n cannot divide both of two consecutive odd integers.

1-11. Suppose $a, b, n \in \mathscr{Z}$ and $|a - b| < |n|$. Prove that n cannot divide both a and b.

1-12. Suppose $a \in \mathscr{Z}$. Prove
  (a) $a|n$ for every $n \in \mathscr{Z}$ if and only if $a = \pm 1$.
  (b) $n|a$ for every $n \in \mathscr{Z}$ if and only if $a = 0$.

1-13. Let $a, b, c \in \mathscr{Z}$ and suppose $c \neq 0$. Show that $ac|bc$ implies $a|b$.

---

* An exercise marked with an asterisk contains a result which will be used in some later discussion.

2. The Greatest Common Divisor and the Least Common Multiple

Definition. *Suppose* $a, b \in \mathscr{Z}$, *and not both of* a *and* b *are zero. The* greatest common divisor (gcd) *of* a *and* b, *denoted* (a,b), *is an integer* d *satisfying*
(a) $d \in \mathscr{Z}^+$,
(b) $d|a$ *and* $d|b$,
(c) *if* $e|a$ *and* $e|b$, *then* $e|d$.

For example, $(4, 12) = 4$, $(2, 3) = 1$, $(6, 8) = 2$. It is not obvious that a gcd of two integers, not both zero, always exists, but this will be proved in Theorem 2.1. However, it is easy to see that if a gcd exists, then it is unique, because if $d = (a,b)$ and $e = (a,b)$, then $e|a$ and $e|b$ by (b); hence, $e|d$ by (c). Similarly $d|e$. But from (a), both d and e are positive, so $d = e$.

Theorem 2.1. *If* a *and* b *are not both zero, then the gcd* (a,b) *exists, is the least positive member of the set*

$$\mathscr{S} = \{au + bv : u, v \in \mathscr{Z}\},$$

and (a,b) *divides every element of* $\mathscr{S}$. *In particular, there exist* $x, y \in \mathscr{Z}$ *such that* $(a,b) = ax + by$.

Proof. Clearly the integers $\pm a = a(\pm 1) + b(0)$ and $\pm b = a(0) + b(\pm 1)$ are all in $\mathscr{S}$ and at least one of $\pm a$ or $\pm b$ is positive, so $\mathscr{S}$ contains positive elements. Therefore, $\mathscr{S}$ contains a least positive integer, say d. Suppose $d = au_0 + bv_0$. We will prove that $d = (a,b)$.

Let $c = au_1 + bv_1$ be any element of $\mathscr{S}$. There exist q, r such that $c = dq + r$, $0 \le r < d$. Now the integer r is in $\mathscr{S}$, since

$$r = c - dq = (au_1 + bv_1) - (au_0 + bv_0)q$$

$$= a(u_1 - u_0 q) + b(v_1 - v_0 q).$$

But $r < d$ implies $r = 0$ since d is the least positive element of $\mathscr{S}$. Therefore, $c = dq$; we have proved that $d|c$ for every $c \in \mathscr{S}$.

Since $a, b \in \mathscr{S}$, we have $d|a$ and $d|b$. Thus, d has the properties (a) and (b) of the gcd, and we now show that d also has property (c). Suppose $e|a$ and $e|b$. By Theorem 1.1(d), e divides any linear combination of a and b. But d is such a combination, so $e|d$, and it follows that $d = (a,b)$. The proof is complete.

We can say a little more about the set $\mathscr{S}$ of Theorem 2.1. We showed above that if $c \in \mathscr{S}$, then $d|c$; the converse is also true, namely, if c is any multiple of d, then $c \in \mathscr{S}$, because if $d|c$, then $dn = c$ for some $n \in \mathscr{Z}$. We have $d = ax + by$ from the theorem, so $c = dn = a(xn) + b(yn) \in \mathscr{S}$. This proves the following corollary.

Corollary 2.1a. *Suppose a, b, c ∈ $\mathscr{Z}$, a and b are not both zero, and d = (a,b). Then there exist integers u and v such that c = au + bv if and only if d|c.*

If $(a,b) = 1$, then a and b are called *relatively prime*. If a and b are relatively prime, then clearly there are integers u and v such that $au + bv = 1$. Conversely, if $au + bv = 1$ for some $u, v \in \mathscr{Z}$, then by Corollary 2.1a, $(a,b)|1$, and since $0 < (a,b)$ we must have $(a,b) = 1$. Therefore we have the following.

Corollary 2.1b. *The relative primality of a and b is necessary and sufficient for the existence of u, v ∈ $\mathscr{Z}$ such that au + bv = 1.*

Some typical facts relating the concepts introduced above are given in the next theorems.

Theorem 2.2. *If $(a,b) = d$, then $(a/d, b/d) = 1$.*

Proof. Let $d = (a,b)$. Then there exist x and y such that $d = ax + by$. Therefore

$$1 = \frac{a}{d}x + \frac{b}{d}y.$$

But a/d and $b/d \in \mathscr{Z}$ so, by Corollary 2.1b, $(a/d, b/d) = 1$.

Theorem 2.3. *If $a|bc$ and $(a,b) = 1$, then $a|c$.*

Proof. From $(a,b) = 1$ we have $ax + by = 1$ for some $x, y \in \mathscr{Z}$. Then $acx + bcy = c$. But $a|ac$ and $a|bc$, so by Theorem 1.1(d), $a|c$.

Definition. *Suppose a, b ∈ $\mathscr{Z}$, ab ≠ 0. The* least common multiple (lcm) *of a and b, denoted $\langle a,b \rangle$, is an integer m defined by*
(a) $m \in \mathscr{Z}^+$,
(b) $a|m$ *and* $b|m$,
(c) *if* $a|n$ *and* $b|n$, *then* $m|n$.

For example, $\langle 4,12 \rangle = 12$, $\langle 2,3 \rangle = 6$, $\langle 6,8 \rangle = 24$. It is easy to see that $\langle a,b \rangle$ always exists for any pair of non-zero integers a and b, and in fact the lcm is the least positive member of the set

$$\mathscr{S} = \{n : n \in \mathscr{Z}^+, a|n, b|n\}.$$

The essential part of the argument should be familiar by now. Since $|ab| \in \mathscr{S}$ there is a least positive member $m \in \mathscr{S}$. Clearly m has properties (a) and (b). Suppose $n \in \mathscr{Z}$ and $a|n, b|n$. We find $q, r \in \mathscr{Z}$ such that $n = mq + r, 0 \le r < m$. Now $a|n, a|m$, so $a|r$. Similarly, $b|r$, so either $r \in \mathscr{S}$ or $r = 0$. By the way m was chosen, $r = 0$. Therefore $m|n$, and m has property (c) so that $m = \langle a,b \rangle$.

There is a fundamental relationship between the gcd and the lcm of two non-zero integers which is given in Theorem 2.4.

Theorem 2.4. *Suppose* $ab \neq 0$. *Then* $(a,b)\langle a,b \rangle = |ab|$.

Proof. Suppose $(a,b) = d$ and $\langle a,b \rangle = m$. Our method of proof will be to show that $ab|dm$ and $dm|ab$. We first observe that $d|ab$ (because $d|a$, so $d|ab$) and that $m|ab$ (since $a|ab$ and $b|ab$, the definition of the lcm gives $m|ab$), or what is the same thing, that $ab/d \in \mathscr{Z}$ and $ab/m \in \mathscr{Z}$.

Now $d|b$, so $ad|ab$, or $a|(ab/d)$. Similarly, $b|(ab/d)$, so $m|(ab/d)$, or $dm|ab$. But $b|m$, so $ab|am$, or $(ab/m)|a$. Similarly, $(ab/m)|b$, so $(ab/m)|d$, or $ab|dm$. Therefore, $dm = \pm ab = |ab|$.

EXERCISES

2-1.*   Prove
     (a) $(ac,bc) = |c|(a,b)$;
     (b)   if $a|c$, $b|c$, and $(a,b) = 1$, then $ab|c$;
     (c) $(a,b) = |a|$ if and only if $a|b$;
     (d)   if $(a,b) = 1$ and $(a,c) = 1$, then $(a,bc) = 1$.

2-2.   Prove
     (a) $\langle ac,bc \rangle = |c| \langle a,b \rangle$;
     (b) $\langle a,b \rangle = |a|$ if and only if $b|a$.

2-3.   Prove
     (a)   if $(a,b) = (a,c)$ and $\langle a,b \rangle = \langle a,c \rangle$, then $b = \pm c$;
     (b) $\langle a,(b,c) \rangle = (\langle a,b \rangle, \langle a,c \rangle)$;
     (c) $(a, \langle b,c \rangle) = \langle (a,b), (a,c) \rangle$.

2-4.*   Suppose not both of a, b are zero. Let

$$\mathscr{D}_1 = \{d : d|a\} \quad \text{and} \quad \mathscr{D}_2 = \{d : d|b\}.$$

Show that $\mathscr{D}_1 \cap \mathscr{D}_2$ is a non-empty finite set. Prove that (a,b) is the largest element in $\mathscr{D}_1 \cap \mathscr{D}_2$.

2-5.   Suppose $f(x)$ is a polynomial of degree ≥ 1 with coefficients in $\mathscr{Z}$. If the equation $f(x) = 0$ has a rational root a/b, $b \neq 0$, $(a,b) = 1$, prove that a divides the constant term of $f(x)$ and b divides the leading coefficient of $f(x)$.

2-6.   Prove this partial converse of Theorem 2.2: If $0 < d$, $d|a$, $d|b$, and $(a/d, b/d) = 1$, then $(a,b) = d$.

2-7.   Suppose $ab \neq 0$. Let

$$\mathscr{E}_1 = \{e : a|e\} \quad \text{and} \quad \mathscr{E}_2 = \{e : b|e\}.$$

Prove that $\langle a,b \rangle$ is the smallest positive element in $\mathscr{E}_1 \cap \mathscr{E}_2$.

2-8.*   The gcd of more than two integers is defined as follows: If $a_1, \ldots, a_n$ ($n \geq 2$) are not all zero, then their gcd $(a_1, \ldots, a_n) = d$ is the unique positive integer such that $d|a_i$ ($i = 1, \ldots, n$), and if $e|a_i$ ($i = 1, \ldots, n$), then $e|d$. Prove by induction that for $n > 2$, $(a_1, \ldots, a_n) = ((a_1, \ldots, a_{n-1}), a_n)$.

2-9.* The lcm of more than two integers all different from zero is the unique positive integer $m = \langle a_1, \ldots, a_n \rangle$ such that $a_i | m$ $(i = 1, \ldots, n)$, and if $a_i | M$ $(i = 1, \ldots, n)$, then $m | M$. Prove that for $n > 2$, $\langle a_1, \ldots, a_n \rangle = \langle \langle a_1, \ldots, a_{n-1} \rangle, a_n \rangle$.

2-10. If $(a,b) = 1$, we know there are integers u and v such that $au + bv = 1$. Prove $(u,v) = (b,u) = (a,v) = 1$.

2-11.* If $b \neq 0$ and $a = bq + r$, $0 \leq r < |b|$, prove $(a,b) = (b,r)$.

2-12. An *integral module* is a set $\mathcal{M}$ such that (a) $\varnothing \neq \mathcal{M} \subset \mathcal{Z}$, and (b) if $a \in \mathcal{M}$, $b \in \mathcal{M}$, then $a + b \in \mathcal{M}$ and $a - b \in \mathcal{M}$. Prove that if $\mathcal{M}$ is any integral module, there exists $m \in \mathcal{Z}$ (m depends on $\mathcal{M}$) such that $\mathcal{M} = \{0, \pm m, \pm 2m, \pm 3m, \ldots\}$.

2-13. Suppose $\mathcal{S}$ is a set such that (a) $\varnothing \neq \mathcal{S} \subset \mathcal{Z}$, and (b) if $a \in \mathcal{S}$, $b \in \mathcal{S}$, then $a - b \in \mathcal{S}$. Prove that $\mathcal{S}$ is an integral module (see Exercise 2-12).

2-14. If $n \in \mathcal{Z}^+$ is fixed, consider the n (not necessarily distinct) integers d_i defined by $d_i = (i,n)$, $i = 1, \ldots, n$. Prove that the set $\{d_1, \ldots, d_n\}$ is just the set $\mathcal{S}$ of Exercise 1-7.

2-15. Give an example to show that the solution x, y of the equation $(a,b) = ax + by$ is not unique.

2-16. Find necessary and sufficient conditions on a and b so that $(a,b) = \langle a,b \rangle$.

2-17. Prove that for every $n \in \mathcal{Z}$, $(3, n^2 + 1) = 1$, $(5, n^2 + 2) = 1$, and $(7, n^2 + 2) = 1$.

2-18. Suppose $\{a_1, a_2, \ldots\}$ is an infinite sequence of integers satisfying $a_{n+2} = a_{n+1} + a_n$ $(n \geq 1)$. If $(a_1, a_2) = 1$, prove $(a_n, a_{n+1}) = 1$ for all $n \in \mathcal{Z}^+$. If $(a_1, a_2) = d$, prove $d | (a_n, a_{n+1})$ for all $n \in \mathcal{Z}^+$.

2-19. If $a, b, n \in \mathcal{Z}^+$ and $n > 1$, find the gcd and the lcm of $n^a - 1$ and $n^b - 1$.

3. The Euclidean Algorithm

We have proved the existence and uniqueness of the gcd of two integers not both zero. Moreover, implicit in Exercise 2-4 is a method for finding the gcd, though that method is inefficient and cumbersome for very large a,b. We now consider a more efficient method for finding the gcd of two integers.

Let us begin with an example. To find $(245,1022)$ we repeatedly use the division algorithm to find

(1) $\qquad 1022 = 245(4) + 42, \qquad 0 \leq 42 < 245$

(2) $\qquad 245 = 42(5) + 35, \qquad 0 \leq 35 < 42$

(3) $\qquad 42 = 35(1) + 7, \qquad 0 \leq 7 < 35$

(4) $\qquad 35 = 7(5) + 0, \qquad 0 \leq 0 < 7.$

Now we can say that $7 = (245,1022)$, because from (4), $7|35$; therefore $7|7$ and $7|35$, so 7 divides any linear combination of 7 and 35. In particular, 42 is such a combination, from (3), so $7|42$. Now $7|35$ and $7|42$, so $7|245$ since (2) shows 245 is a linear combination of 35 and 42. Finally, we see from (1) that $7|1022$ since $7|42$ and $7|245$. Hence, 7 is a common divisor of 245 and of 1022.

To show that 7 is the greatest of the common divisors, suppose $e|245$ and $e|1022$. First we rewrite (1), (2), and (3) in the form

(5) $$42 = 1022 + 245(-4)$$

(6) $$35 = 245 + 42(-5)$$

(7) $$7 = 42 + 35(-1).$$

From (5) we see that $e|42$; then from (6) we see that $e|35$. Consequently, (7) tells us that $e|7$, so that 7 is the gcd. This entire argument may be given more elegantly by using Exercise 2-11 and Exercise 2-1(c) to observe that $(1022,245) = (245,42) = (42,35) = (35,7) = 7$.

This process for finding the gcd is known as the *Euclidean algorithm*. We can also use this algorithm to find integers x and y such that $7 = 1022x + 245y$, as follows. Substitute from (6) into (7) to get

$$7 = 42 + 35(-1)$$
$$= 42 + (245 + 42(-5))(-1)$$
$$= 245(-1) + 42(6);$$

now substitute from (5) to get

$$7 = 245(-1) + (1022 + 245(-4))(6)$$
$$= 1022(6) + 245(-25).$$

The above algorithm may be applied to any pair of integers a,b not both zero. Since $(a,0) = |a|$ for every $a \neq 0$, and since $(a,b) = (|a|,|b|)$, we may assume that both a and b are positive. Then we use the division algorithm repeatedly to find q_i, r_i $(i = 1, \ldots, k + 1)$ such that

$$a = bq_1 + r_1, \qquad 0 < r_1 < b$$
$$b = r_1 q_2 + r_2, \qquad 0 < r_2 < r_1$$
$$r_1 = r_2 q_3 + r_3, \qquad 0 < r_3 < r_2$$
$$\cdots$$
$$r_{k-2} = r_{k-1} q_k + r_k, \qquad 0 < r_k < r_{k-1}$$
$$r_{k-1} = r_k q_{k+1} + r_{k+1}, \qquad 0 = r_{k+1} < r_k.$$

Then $(a,b) = r_k$, the last non-zero remainder. It is clear that the process must terminate with a remainder zero after a finite number of steps, because $r_1 > r_2 > \cdots \geq 0$ is a *decreasing* sequence of non-negative integers and r_{k+1} must be zero for some k.

Also, as in the example, the sequence of equations can be used to find x,y such that $r_k = ax + by$. The lcm of two integers can be found by using Theorem 2.4. In the exercises below the student will show that the gcd of more than two integers can be found by using Exercise 2-8 and the Euclidean algorithm; the lcm can be found by using Exercise 2-9 and Theorem 2.4.

EXERCISES

3-1. Find $(20,35)$, $(112,96)$, $(27,45)$.

3-2. Use the Euclidean algorithm to find integers x,y such that $(20,35) = 20x + 35y$; $(112,96) = 112x + 96y$; $(27,45) = 27x + 45y$.

3-3. Find $\langle 20,35 \rangle$, $\langle 112,96 \rangle$, $\langle 27,45 \rangle$.

3-4. Use Exercise 2-8 and the Euclidean algorithm to find
   (a) $(60,30,42,8)$,
   (b) $(42,60,8,30)$,
   (c) $(2250,30,540,900)$.

3-5. Use Exercise 2-9 and Theorem 2.4 to find
   (a) $\langle 60,30,42,8 \rangle$,
   (b) $\langle 42,60,8,30 \rangle$,
   (c) $\langle 2250,30,540,900 \rangle$.

3-6. Suppose k is a non-zero integer and m and n are distinct positive integers. Prove that

$$(2k)^{2^n} + 1 \quad \text{and} \quad (2k)^{2^m} + 1$$

are relatively prime.

4. The Fundamental Theorem

Definition. *A positive integer p is called a* prime *if and only if p has exactly two (distinct) positive divisors. An integer $n > 1$ which is not prime is called* composite.

We note that if $n \in \mathscr{Z}$ and p is prime, then (n,p) is either 1 or p, and the latter occurs if and only if $p|n$.

Theorem 4.1. *Suppose p is prime and $a_1, \ldots, a_k \in \mathscr{Z}$. If $p|a_1 \cdots a_k$, then $p|a_i$ for some i, $1 \leq i \leq k$.*

Proof. We prove this theorem by induction on k. For $k = 1$ there is nothing to prove. For $k = 2$, assume $p|a_1 a_2$; if $p|a_1$ we are through, so suppose $p \nmid a_1$. Then $(p,a_1) = 1$, and by Theorem 2.3 we have $p|a_2$.

Now assume the theorem has been proved for $k = 1, \ldots, n - 1$ $(2 \le n)$ and suppose $p | a_1 \cdots a_n$. We have $p | (a_1 \cdots a_{n-1}) a_n$; therefore, either $p | a_n$ or $p | a_1 \cdots a_{n-1}$. If $p | a_n$ we are through. If not, then $p | a_1 \cdots a_{n-1}$, and by the inductive assumption, $p | a_i$ for some i, $1 \le i \le n - 1$. In any case, then, if $p | a_1 \cdots a_n$, then $p | a_i$ for some i, $1 \le i \le n$, and the proof is complete.

Theorem 4.2. (*Fundamental theorem of number theory.*) *If $n > 1$, then n can be represented as a product of a finite number of primes, and this representation is unique except for the order of the prime factors.*

Proof. (Existence.) For $n = 2$, we obviously have n expressed as a product of primes. Suppose for all n, $2 \le n \le N - 1$, it has been shown that n is a product of a finite number of primes. If N is itself a prime, we are through. If N is not a prime, then it has a divisor d, $1 < d < N$, say $de = N$. But then $1 < e < N$, and the inductive hypothesis guarantees that both d and e have a prime factor decomposition, and hence so does N.

(Uniqueness.) We again proceed by induction. For $n = 2$, the factorization is unique. Suppose uniqueness has been proved for all n, $2 \le n < N$. If N is prime, its factorization is unique, so suppose N is composite. We will assume that N has at least two representations as a product of primes and reach a contradiction.

Suppose $N = p_1 \cdots p_r$ and $N = q_1 \cdots q_s$, where p_i $(i = 1, \ldots, r)$ and q_j $(j = 1, \ldots, s)$ are primes. Clearly r and s are both larger than 1 since N is not prime. From

$$p_1 \cdots p_r = N = q_1 \cdots q_s$$

we have $p_1 | q_1 \cdots q_s$, so $p_1 | q_k$ for some k, $1 \le k \le s$ by Theorem 4.1. Since $1 < p_1$, and q_k has exactly two positive divisors, we know $p_1 = q_k$.

But then $N/p_1 < N$ and $1 < p_2 \cdots p_r = N/p_1$, so that $N/p_1 = N/q_k$ has a unique representation as a product of primes. Therefore, $r = s$, and for each $i = 2, \ldots, r$, it follows that $p_i = q_j$ for some j, $j \ne k$, $1 \le j \le s$. Since $N/p_1 = p_2 \cdots p_r$ uniquely, then $N = p_1 p_2 \cdots p_r$ is the unique factorization of N. This completes the induction.

From the factorization of $n > 1$ into primes, we obtain a representation

(8)
$$n = \prod_{i=1}^{r} p_i^{\alpha_i}$$

where the p_i are distinct primes and the $\alpha_i \in \mathscr{Z}^+$. Often it is convenient to allow $\alpha_i \ge 0$. The factorization (8) is called the *canonical* form of n. For example, $15 = 3^1 \cdot 5^1$ and $25 = 5^2$, though sometimes it might be desirable to write $25 = 3^0 \cdot 5^2$. Thus we could say that the canonical forms for 15 and 25 formally involve the same primes.

EXERCISES

4-1. Suppose

$$1 < n = \prod_{i=1}^{r} p_i^{\alpha_i}, \quad \alpha_i > 0.$$

Prove that for $d > 0$, $d|n$ if and only if

$$d = \prod_{i=1}^{r} p_i^{\beta_i}$$

with $0 \le \beta_i \le \alpha_i$ for $i = 1, \ldots, r$.

4-2.* Suppose

$$a = \prod_{i=1}^{r} p_i^{\alpha_i}, \quad \alpha_i \ge 0, \quad \text{and} \quad b = \prod_{i=1}^{r} p_i^{\beta_i}, \quad \beta_i \ge 0.$$

Prove that

$$(a,b) = \prod_{i=1}^{r} p_i^{\gamma_i}$$

where $\gamma_i = \min(\alpha_i, \beta_i)$, $i = 1, \ldots, r$. Hence, use Theorem 2.4 to conclude that

$$\langle a,b \rangle = \prod_{i=1}^{r} p_i^{\delta_i}$$

where $\delta_i = \max(\alpha_i, \beta_i)$.

4-3.* If $(m,n) = 1$ and $0 < d|mn$, prove that d has a unique factorization $d = ab$ such that $a|m$, $b|n$, and $(a,b) = 1$ and $(m/a, n/b) = 1$. Prove that $a = (d,m)$ and $b = (d,n)$.

4-4. Suppose $k > 2$ and $a_1, \ldots, a_k \in \mathscr{Z}^+$. If

$$a_j = \prod_{i=1}^{r} p_i^{\alpha_{ji}}, \quad \alpha_{ji} \ge 0 \quad (j = 1, \ldots, k),$$

prove that

$$(a_1, \ldots, a_k) = \prod_{i=1}^{r} p_i^{\beta_i}$$

where $\beta_i = \min(\alpha_{1i}, \ldots, \alpha_{ki})$ for $i = 1, \ldots, r$.

4-5. With the same notation as in Exercise 4-4, prove that

$$\langle a_1, \ldots, a_k \rangle = \prod_{i=1}^{r} p_i^{\delta_i}$$

where $\delta_i = \max(\alpha_{1i}, \ldots, \alpha_{ki})$ for $i = 1, \ldots, r$.

4-6. Notice that $(2,4,6) = 2$ and that $\langle 2,4,6 \rangle = 12$, so that $24 = (2,4,6) \times \langle 2,4,6 \rangle \neq 2 \cdot 4 \cdot 6$. Thus, Theorem 2.4 cannot be extended without some restrictions. Find necessary and sufficient conditions so that for $k > 2$,

$$(a_1, \ldots, a_k)\langle a_1, \ldots, a_k \rangle = |a_1 \cdots a_k|.$$

4-7.* Suppose $p_1, \ldots, p_n$ are primes and let $m = p_1 \cdots p_n + 1$. Prove that m has a prime divisor p such that $p \neq p_1, \ldots, p \neq p_n$. Hence prove that there are infinitely many primes.

4-8. Suppose $\mathscr{S}$ is a set containing $n + 1$ integers t such that $1 \leq t \leq 2n$ for every $t \in \mathscr{S}$. Prove there exist integers $a, b \in \mathscr{S}$, $a \neq b$, such that $a|b$. ["Elementary Problem E1765," *Amer. Math. Monthly*, Vol. 72 (1965), p. 183.]

4-9. Use the results of Exercise 4-2 to prove
(a) $\langle a, (b,c) \rangle = (\langle a,b \rangle, \langle a,c \rangle)$,
(b) $(a, \langle b,c \rangle) = \langle (a,b), (a,c) \rangle$.

4-10. Let $\mathscr{S} = \{3n - 2 : n \in \mathscr{Z}^+\}$. Define a *prime in* $\mathscr{S}$ to be any element of $\mathscr{S}$ which has exactly two distinct divisors in $\mathscr{S}$. For example, 4 and 7 are primes in $\mathscr{S}$.
(a) Prove that the product of two elements in $\mathscr{S}$ is in $\mathscr{S}$.
(b) Given any $t \in \mathscr{S}$, prove that either $t = 1$, t is a prime in $\mathscr{S}$, or t can be written as a product of primes in $\mathscr{S}$.
(c) Show that 10 and 25 are primes in $\mathscr{S}$.
(d) Notice that $4 \cdot 25 = 10 \cdot 10$, so that factorization into primes in $\mathscr{S}$ is not unique. Find another element of $\mathscr{S}$ which has at least two different factorizations into primes in $\mathscr{S}$.

4-11. Consider the table

4	7	10	13	16	19	$\cdots$
7	12	17	22	27	32	$\cdots$
10	17	24	31	38	45	$\cdots$
13	22	31	40	49	58	$\cdots$

. . . .

in which the rule of formation is evident. Show that a positive integer k occurs in the table if and only if $2k + 1$ is not a prime.

4-12. Find a prime which is simultaneously of each of the forms $a^2 + b^2$, $a^2 + 2b^2, \ldots, a^2 + 10b^2$ for some $a, b \in \mathscr{Z}$.

Chapter 2

ARITHMETIC FUNCTIONS

5. The Semigroup $\mathscr{A}$ of Arithmetic Functions

Definition. *An* arithmetic function *is a function whose domain is $\mathscr{Z}^+$ and whose range is a subset of the set of complex numbers.*

For example, $f(n) = n$ for all $n \in \mathscr{Z}^+$, and $g(n) = e^{in}$ where $i^2 = -1$ are arithmetic functions. Let $\mathscr{A}$ denote the set of all arithmetic functions. If f and $g \in \mathscr{A}$, the *product* (or *Dirichlet convolution product*) of f and g is the function denoted by $f \cdot g$ and defined at each $n \in \mathscr{Z}^+$ by

$$f \cdot g(n) = \sum_{d|n} f(d)\, g\left(\frac{n}{d}\right)$$

where the summation extends over all positive divisors d of n.

Examples.

$$f \cdot g(1) = f(1)\, g(1)$$

$$f \cdot g(2) = f(1)\, g(2) + f(2)\, g(1)$$

$$f \cdot g(6) = f(1)\, g(6) + f(2)\, g(3) + f(3)\, g(2) + f(6)\, g(1)$$

Notice that this product is in fact a binary operation on $\mathscr{A}$ since $f, g \in \mathscr{A}$ implies $f \cdot g \in \mathscr{A}$, which follows from the fact that sums and products of complex numbers are complex numbers. It will be convenient to have the following equivalent form of the definition of $f \cdot g$.

Lemma 5.1. *If $f, g \in \mathscr{A}$, then for every $n \in \mathscr{Z}^+$,*

$$f \cdot g(n) = \sum_{ab=n} f(a)\, g(b)$$

where the summation extends over all ordered pairs a, b of positive integers whose product is n.

13

Proof. By Exercise 1-7, $d|n$ if and only if $(n/d)|n$. Therefore, as d ranges over all positive divisors of n, the ordered pairs d, n/d range over all ordered pairs of positive integers whose product is n.

Theorem 5.1. *The product of arithmetic functions is associative; that is, if $f, g, h \in \mathscr{A}$, then $(f \cdot g) \cdot h(n) = f \cdot (g \cdot h)(n)$ for every $n \in \mathscr{Z}^+$.*
Proof. For any fixed $n \in \mathscr{Z}^+$,

$$
\begin{aligned}
(f \cdot g) \cdot h(n) &= \sum_{dc=n} f \cdot g(d)\, h(c) \\
&= \sum_{dc=n} \left\{ \sum_{ab=d} f(a)\, g(b) \right\} h(c) \\
&= \sum_{abc=n} f(a)\, g(b)\, h(c) \\
&= \sum_{ae=n} f(a) \left\{ \sum_{bc=e} g(b)\, h(c) \right\} \\
&= \sum_{ae=n} f(a)\, g \cdot h(e) \\
&= f \cdot (g \cdot h)(n).
\end{aligned}
$$

An algebraic system $(\mathscr{S}, \times)$ consisting of a set $\mathscr{S}$ together with an operation " $\times$ " on $\mathscr{S}$ is called a *semigroup* provided that " $\times$ " is associative. In this terminology we have proved that $(\mathscr{A}, \cdot)$ is a semigroup. This semigroup of arithmetic functions has other interesting properties, as we now show.

Theorem 5.2. *The product of arithmetic functions is a commutative operation; that is, if $f, g \in \mathscr{A}$, then $f \cdot g(n) = g \cdot f(n)$ for all $n \in \mathscr{Z}^+$.*
Proof.

$$
\begin{aligned}
f \cdot g(n) &= \sum_{ab=n} f(a)\, g(b) \\
&= \sum_{ba=n} g(b)\, f(a) \\
&= g \cdot f(n).
\end{aligned}
$$

We now consider this question: Does there exist a function ε in $\mathscr{A}$ such that $f \cdot \varepsilon = \varepsilon \cdot f = f$ for *every* $f \in \mathscr{A}$? The reader is probably aware of the fact that such an element in a semigroup is called the *identity* of the semigroup. Thus, our question is: Does there exist an identity ε in $(\mathscr{A}, \cdot)$? Evidently, since the product of functions is commutative, it will suffice to investigate the existence of a function ε such that $f = \varepsilon \cdot f$ for all $f \in \mathscr{A}$, or equivalently, such that for every $f \in \mathscr{A}$,

(1) $$ f(n) = \varepsilon \cdot f(n) \quad \text{for all} \quad n \in \mathscr{Z}^+. $$

We shall assume there is such an identity ε and find the necessary properties which this function must have. Writing out (1) for $n = 1$, we have $f(1) = \varepsilon(1)f(1)$. Since this last equation must hold for all $f \in \mathscr{A}$, it must hold in particular if $f(1) \neq 0$, so we must have $\varepsilon(1) = 1$. Writing out (1) for $n = 2$ gives

$$f(2) = \varepsilon(1)f(2) + \varepsilon(2)f(1).$$

Since $\varepsilon(1) = 1$, the above equation reduces to $0 = \varepsilon(2)f(1)$, which will hold for arbitrary f only if $\varepsilon(2) = 0$.

Now we continue by induction. Assume that $\varepsilon(n) = 0$ for $n = 2, \ldots, k - 1$ ($k \geq 3$) and consider (1) for $n = k$.

$$f(k) = \sum_{d \mid k} \varepsilon(d) f\left(\frac{k}{d}\right)$$

$$f(k) = \varepsilon(1) f(k) + \sum_{\substack{d \mid k \\ 1 < d < k}} \varepsilon(d) f\left(\frac{k}{d}\right) + \varepsilon(k) f(1)$$

Any terms appearing in the summation will vanish by the inductive assumption; also $\varepsilon(1) = 1$, so the above equation is simply $0 = \varepsilon(k)f(1)$. Thus, $\varepsilon(k) = 0$ is necessary, and by induction we have: If ε is the identity in $(\mathscr{A}, \cdot)$, then $\varepsilon(1) = 1$ and $\varepsilon(n) = 0$ if $n > 1$. The converse of this statement answers the question asked earlier. The proof of the converse is left as an exercise.

A convenient description of the function ε can be given in terms of the *largest integer* function (sometimes called the "square bracket" function). If x is a real number, then $[x]$ is defined by

$$[x] = \max \{n : n \in \mathscr{Z}, n \leq x\}.$$

Thus, $[x]$ is the largest integer which does not exceed x; for example, $[2] = 2$, $[\sqrt{3}] = 1$, $[-7/2] = -4$. Now notice that $\varepsilon(n) = [1/n]$.

The results of this section are summarized in Theorem 5.3.

Theorem 5.3. *The system $(\mathscr{A}, \cdot)$ is a commutative semigroup with identity ε, where ε is the arithmetic function satisfying $\varepsilon(n) = [1/n]$.*

EXERCISES

5-1.* Prove that ε is the identity in $(\mathscr{A}, \cdot)$.

5-2. Suppose $f(n) = 1$ for all $n \in \mathscr{Z}^+$, and suppose p is a prime. For $k \in \mathscr{Z}^+$, find $f \cdot f(p^k)$.

5-3. Suppose $f(2) = -1$ and $f(n) = 0$ if $n \neq 2$, $g(3) = 1$ and $g(n) = 0$ if $n \neq 3$. Find $f \cdot g$.

5-4.   Suppose $f(n) = 1$ for every $n \in \mathscr{Z}^+$, $g(1) = 1$, $g(p) = -1$, and $g(p^k) = 0$ for $k > 1$, where p is a fixed prime. Find $f \cdot f \cdot g(p^k)$ for all $k \geq 0$.

5-5.* Define $\theta(n) = 0$ for every $n \in \mathscr{Z}^+$. Show that $\theta \cdot f = \theta$ for every $f \in \mathscr{A}$, and if $g \cdot f = g$ for every $f \in \mathscr{A}$, then $g = \theta$.

5-6.   An element f in a semigroup is called an *idempotent* if it has the property that $f \cdot f = f$. Show that ε and θ are the only idempotents in $\mathscr{A}$.

6. The Group of Units in $\mathscr{A}$

Definition. *Suppose $f \in \mathscr{A}$. If there is a function $f' \in \mathscr{A}$ such that $f \cdot f' = \varepsilon$, then f' is called the* inverse *of f.*

Evidently not every arithmetic function has an inverse. For example, there can be no function θ' (see Exercise 5-5 for the definition of θ) such that $\theta \cdot \theta'(n) = \varepsilon(n)$ for every n, because at $n = 1$ this would require that

$$0 = \theta(1)\,\theta'(1) = \theta \cdot \theta'(1) = \varepsilon(1) = 1,$$

which is impossible. The example shows essentially the only way that a function can fail to have an inverse. The situation is described in the next theorem.

Theorem 6.1. *A necessary and sufficient condition that the inverse of f exist is that $f(1) \neq 0$.*

Proof. Suppose f' exists. Since $f \cdot f'(1) = \varepsilon(1)$, we have $f(1) \neq 0$.

Conversely, suppose $f(1) \neq 0$ and consider the function g defined inductively by

$$g(1) = \frac{1}{f(1)}$$

$$g(n) = \frac{-1}{f(1)} \sum_{\substack{cd=n \\ 1 < c}} f(c)\,g(d), \qquad n > 1.$$

We will show that $f \cdot g = \varepsilon$, which will prove that $g = f'$ (see Exercise 6-1). Obviously $f \cdot g(1) = 1 = \varepsilon(1)$ and

$$f \cdot g(2) = f(1)\,g(2) + f(2)\,g(1)$$

$$= f(1)\left\{ \frac{-1}{f(1)} f(2)\,g(1) \right\} + f(2)\,g(1)$$

$$= 0 = \varepsilon(2).$$

Suppose $f \cdot g(k) = 0$ for $k = 2, \ldots, n - 1$ $(n \geq 3)$. Then

$$f \cdot g(n) = f(n) g(1) + \sum_{\substack{ab=n \\ 1<b}} f(a) g(b)$$

$$= f(n) g(1) + \sum_{\substack{ab=n \\ 1<b}} f(a) \left\{ \frac{-1}{f(1)} \sum_{\substack{cd=b \\ 1<c}} f(c) g(d) \right\}$$

$$= f(n) g(1) - \frac{1}{f(1)} \sum_{\substack{ab=n \\ 1<b}} f(a) \left\{ \sum_{cd=b} f(c) g(d) - f(1) g(b) \right\}$$

$$= f(n) g(1) - \frac{1}{f(1)} \sum_{\substack{ab=n \\ 1<b}} f(a) \sum_{cd=b} f(c) g(d) + \sum_{\substack{ab=n \\ 1<b}} f(a) g(b)$$

$$= f(n) g(1) - \frac{1}{f(1)} \sum_{\substack{ab=n \\ 1<b}} f(a) f \cdot g(b) + \sum_{\substack{ab=n \\ 1<b}} f(a) g(b).$$

By the inductive assumption, the only non-zero term in the first summation occurs for $b = n$, so the above simplifies to

$$f \cdot g(n) = f(n) g(1) - \frac{1}{f(1)} f(1) f \cdot g(n) + \sum_{\substack{ab=n \\ 1<b}} f(a) g(b) = 0.$$

This completes the induction and shows that for all n, $f \cdot g(n) = \varepsilon(n)$.

Definition. *The set* $\mathscr{U} \subset \mathscr{A}$ *is defined by*

$$\mathscr{U} = \{ f \in \mathscr{A} : f(1) \neq 0 \}.$$

The last theorem shows that $\mathscr{U}$ is just the set of functions which have inverses. This set $\mathscr{U}$ is called the *set of units* in $\mathscr{A}$. It is known that the set of units in a semigroup is a group called the "group of units"; though this result holds in more general situations, we state it only for $\mathscr{U}$. The proof is left as an exercise.

Theorem 6.2. *The set* $\mathscr{U}$ *of units in* $\mathscr{A}$ *is a commutative group. That is,*
(a) *if* $f, g \in \mathscr{U}$, *then* $f \cdot g \in \mathscr{U}$;
(b) *the product of functions in* $\mathscr{U}$ *is associative and commutative*;
(c) *the identity* $\varepsilon \in \mathscr{U}$;
(d) *if* $f \in \mathscr{U}$, *then* f' *exists and* $f' \in \mathscr{U}$.

EXERCISES
6-1. Suppose $f \in \mathscr{A}$. If there exists a pair of functions f' and $g \in \mathscr{A}$ such that $f \cdot f' = \varepsilon$ and $f \cdot g = \varepsilon$, prove that $f' = g$.
6-2. Prove Theorem 6.2.

6-3. If $f(n) = 1$ for all n, find $f'(n)$ for $1 \leq n \leq 10$.
6-4. If $f(n) = 1$ for all n and p is any prime, find $f'(p)$ and $f'(p^2)$.
6-5. Prove that $\varepsilon' = \varepsilon$.
6-6. If f' exists, prove that $(f')'$, the inverse of f', exists and that $(f')' = f$.

7. The Subgroup of Multiplicative Functions

Definition. *Let* $\mathscr{M}' = \{f \in \mathscr{A} : \text{if } (m, n) = 1, \text{ then } f(mn) = f(m)f(n)\}$ *and* $\mathscr{M} = \{f \in \mathscr{M}'; f(n) \neq 0 \text{ for some } n \in \mathscr{Z}^+\}$. *A function* f *is called* multiplicative *if and only if* $f \in \mathscr{M}$.

It is clear that $\mathscr{M}' = \mathscr{M} \cup \{\theta\}$ where $\theta \in \mathscr{A}$ is the function defined in Exercise 5-5. We also note that $f \in \mathscr{M}$ if and only if $f \in \mathscr{M}'$ and $f(1) = 1$, because if $f \in \mathscr{M}'$ and $f(1) = 1 \neq 0$, clearly $f \in \mathscr{M}$; conversely, suppose $f \in \mathscr{M}$ and let $n_0 \in \mathscr{Z}^+$ be such that $f(n_0) \neq 0$. Then $f \in \mathscr{M}'$, and since $(1, n_0) = 1$, $f(1 \cdot n_0) = f(1) f(n_0)$ implies $f(1) = 1$.

From the above remarks it is obvious that $\mathscr{M} \subset \mathscr{U}$. The set $\mathscr{M}$ of multiplicative functions is in fact a subgroup of the group of units, as we now show.

Theorem 7.1. *If* $f, g \in \mathscr{M}$, *then* $f \cdot g \in \mathscr{M}$.

Proof. (In this proof we use the result of Exercise 4-3.) If $f, g \in \mathscr{M}$, then $f(1) = 1 = g(1)$ and it follows that $f \cdot g(1) = 1$. If $(m, n) = 1$, then

$$
\begin{aligned}
f \cdot g(mn) &= \sum_{d \mid mn} f(d) g\left(\frac{mn}{d}\right) \\
&= \sum_{\substack{a \mid m \\ b \mid n \\ ab = d}} f(ab) g\left(\frac{m}{a} \frac{n}{b}\right) \\
&= \sum_{\substack{a \mid m \\ b \mid n}} f(a) f(b) g\left(\frac{m}{a}\right) g\left(\frac{n}{b}\right) \\
&= \left\{\sum_{a \mid m} f(a) g\left(\frac{m}{a}\right)\right\} \left\{\sum_{b \mid n} f(b) g\left(\frac{n}{b}\right)\right\} \\
&= f \cdot g(m) f \cdot g(n).
\end{aligned}
$$

Thus, $f \cdot g \in \mathscr{M}$.

The above theorem may be applied to obtain an interesting corollary. We first introduce a special class of multiplicative functions.

Definition. *Let* s *be any complex number. The functions* $\iota_s \in \mathscr{A}$ *are defined by*

$$\iota_s(n) = n^s, \quad n \in \mathscr{Z}^+.$$

We will refer to these functions as the iota *functions.*

We have left as an exercise the demonstration that every iota function is multiplicative—in particular, the function $\iota_0 \in \mathscr{M}$ [$\iota_0(n) = 1$ for all n]; it follows from Theorem 7.1 that if $f \in \mathscr{M}$, then $f \cdot \iota_0 \in \mathscr{M}$. This is equivalent to the following important result.

Corollary 7.1. *If f is a multiplicative function and if*

$$g(n) = \sum_{d|n} f(d),$$

then g is also multiplicative.

Continuing now with the demonstration that $\mathscr{M}$ is a group, we establish the following.

Theorem 7.2. *If $f \in \mathscr{M}$, then f' exists and $f' \in \mathscr{M}$.*

Proof. If $f \in \mathscr{M} \subset \mathscr{U}$, clearly f' exists. Also $f'(1) = 1/f(1) = 1$. Suppose we have shown that $f'(st) = f'(s)f'(t)$ whenever $(s,t) = 1$ for all s,t such that $1 \le st \le k - 1$. Let $k = mn$ be any factorization of k such that $(m,n) = 1$. Now consider

$$0 = \varepsilon(k) = \sum_{d|k} f(d)\, f'\!\left(\frac{k}{d}\right)$$

$$= \sum_{\substack{a|m \\ b|n}} f(ab)\, f'\!\left(\frac{m}{a}\frac{n}{b}\right)$$

$$= \sum_{\substack{a|m \\ b|n \\ ab>1}} f(ab)\, f'\!\left(\frac{m}{a}\frac{n}{b}\right) + f(1)\, f'(mn)$$

$$= \sum_{\substack{a|m \\ b|n \\ ab>1}} f(a)\, f(b)\, f'\!\left(\frac{m}{a}\right) f'\!\left(\frac{n}{b}\right) + f'(mn)$$

$$= \sum_{\substack{a|m \\ b|n}} f(a)\, f(b)\, f'\!\left(\frac{m}{a}\right) f'\!\left(\frac{n}{b}\right) - f'(m)\, f'(n) + f'(mn)$$

$$= \left\{\sum_{a|m} f(a)\, f'\!\left(\frac{m}{a}\right)\right\}\left\{\sum_{b|n} f(b)\, f'\!\left(\frac{n}{b}\right)\right\} - f'(m)\, f'(n) + f'(mn)$$

(2) $$0 = \varepsilon(m)\, \varepsilon(n) - f'(m)\, f'(n) + f'(mn).$$

Since $1 < k$, at least one of m,n is larger than 1, so $\varepsilon(m)\, \varepsilon(n)$ vanishes. From (2) we have $f'(m)\, f'(n) = f'(mn)$, and by induction, $f' \in \mathscr{M}$.

We have already shown that the product of functions is associative and commutative in $\mathscr{A}$; therefore, the product also has these properties in $\mathscr{M}$. It is easy to see that the identity $\varepsilon \in \mathscr{M}$. These observations, together with our last two theorems, prove Theorem 7.3.

Theorem 7.3. *The set $\mathscr{M}$ of multiplicative functions is a commutative group.*

EXERCISES

7-1.   Show that ε is multiplicative.

7-2.* If $n \in \mathscr{Z}^+$ and s is a complex number, say $s = \sigma + it$ with σ, t real, then $n^s = n^\sigma \{\cos(t \log n) + i \sin(t \log n)\}$. If $m, n \in \mathscr{Z}^+$ and s is complex, prove that $m^s n^s = (mn)^s$. Hence prove that for every complex s, $\iota_s \in \mathscr{M}$.

7-3.   Give an example of a function f such that $f(1) = 1$ but f is not multiplicative.

8. The Möbius Function and Inversion Formulas

The next theorem is of help in studying particular functions in $\mathscr{M}$.

Theorem 8.1. *Suppose $f \in \mathscr{M}$ and $n > 1$ has the canonical form*

$$(3) \qquad n = \prod_{i=1}^{k} p_i^{\beta_i}, \qquad \beta_i > 0.$$

Then

$$f(n) = \prod_{i=1}^{k} f(p_i^{\beta_i}).$$

Proof. Let f and n be as in the hypothesis. We use induction on k. For $k = 1$ there is nothing to prove, so assume the theorem is true for $1 \leq k < K$. Then

$$f\left(\prod_{i=1}^{K} p_i^{\beta_i}\right) = f\left(\left\{\prod_{i=1}^{K-1} p_i^{\beta_i}\right\} p_K^{\beta_K}\right)$$

$$= f\left(\prod_{i=1}^{K-1} p_i^{\beta_i}\right) f(p_K^{\beta_K})$$

$$= \left\{\prod_{i=1}^{K-1} f(p_i^{\beta_i})\right\} f(p_K^{\beta_K})$$

$$= \prod_{i=1}^{K} f(p_i^{\beta_i}).$$

This theorem tells us that a multiplicative function is completely determined by its values at p^β for every prime p and for every $\beta \in \mathscr{Z}^+$. We will show an application of Theorem 8.1 by finding the inverse of $\iota_0(n) = n^0 = 1$ for every n. Since $\iota_0 \in \mathscr{M}$, we know by Theorem 7.2 that ι_0' exists and $\iota_0' \in \mathscr{M}$, so we need only find $\iota_0'(p^\beta)$ for primes p and for $\beta = 1, 2, \ldots$. Therefore, let p be any fixed prime; we proceed by induction on β. For $\beta = 1$ we have

$$0 = \varepsilon(p) = \iota_0' \cdot \iota_0(p)$$

$$= \iota_0'(1)\,\iota_0(p) + \iota_0'(p)\,\iota_0(1)$$

$$= 1 + \iota_0'(p),$$

from which we see that $\iota_0'(p) = -1$.

For $\beta = 2$, $\varepsilon(p^2) = \iota_0' \cdot \iota_0(p^2)$ yields

$$0 = \sum_{j=0}^{2} \iota_0'(p^j)\,\iota_0(p^{2-j}) = 1 - 1 + \iota_0'(p^2),$$

so $\iota_0'(p^2) = 0$. Now assume that $\iota_0'(p^\beta) = 0$ for $2 \le \beta < B$; then

$$0 = \varepsilon(p^B) = \sum_{j=0}^{B} \iota_0'(p^j)$$

$$= \iota_0'(1) + \iota_0'(p) + \sum_{j=2}^{B-1} \iota_0'(p^j) + \iota_0'(p^B)$$

$$= \iota_0'(p^B).$$

We have shown that $\iota_0'(p) = -1$ and $\iota_0'(p^\beta) = 0$ if $\beta \ge 2$ for every prime p.

Traditionally this function ι_0' is called the *Möbius function*, and is denoted by μ. Adopting this notation, we apply Throrem 8.1 to conclude that if $n > 1$ is factored as in (3), then

$$\mu(n) = \prod_{i=1}^{k} \mu(p_i^{\beta_i}),$$

which is 0 if any of the β_i is at least 2, or $(-1)^k$ if every $\beta_i = 1$.

The discussion is summarized in Theorem 8.2.

Theorem 8.2. *The Möbius function* μ, *defined by* $\iota_0 \cdot \mu = \mu \cdot \iota_0 = \varepsilon$, *is a multiplicative function with values given by*

$$\mu(n) = \begin{cases} 1 & \text{if } n = 1; \\ (-1)^k & \text{if } n \text{ is the product of } k \text{ distinct primes}; \\ 0 & \text{if } n \text{ is divisible by the square of some prime}. \end{cases}$$

Equivalently, the Möbius function satisfies the equations $\mu(1) = 1$, *and for* $n > 1$,

$$\sum_{d|n} \mu(d) = 0.$$

Corollary 8.2a. (*Möbius inversion formula.*) *If*

(4) $$g(n) = \sum_{d|n} f(d),$$

then

(5) $$f(n) = \sum_{d|n} g(d) \, \mu\left(\frac{n}{d}\right).$$

Conversely, (5) *implies* (4).

Proof. The equation (4) is simply $g = f \cdot \iota_0$, and (5) is $f = g \cdot \mu$. Evidently $g = f \cdot \iota_0$ if and only if $g \cdot \mu = f \cdot \iota_0 \cdot \mu = f \cdot \varepsilon = f$.

Corollary 8.2b. *Suppose*

$$g(n) = \sum_{d|n} f(d).$$

If g is multiplicative, then so is f.

Proof. This corollary is the converse of Corollary 7.1. If $g = f \cdot \iota_0$, then $g \cdot \mu = f$. Now g and $\mu \in \mathcal{M}$, so their product $f \in \mathcal{M}$.

The Möbius inversion formula suggests the derivation of a wide class of inversion formulas, which we make in our next theorem. The proof of the theorem is trivial. Notice that the Möbius inversion formula is a special case of the theorem.

Theorem 8.3. *Suppose* $h \in \mathcal{U}$ *and* $f, g \in \mathcal{A}$. *Then* $f \cdot h = g$ *if and only if* $f = g \cdot h'$.

EXERCISES

8-1.   Prove Theorem 8.3.

8-2.   An integer n is called *squarefree* if for every $t > 1$, $t^2 \nmid n$.

    (a)   Prove that the integer 1 is the only positive integer which is both squarefree and a square.

    (b)   Show that n is squarefree if and only if $\mu(n) \neq 0$.

    (c)   A function $f \in \mathcal{A}$ is called the *characteristic function* of a set $\mathcal{S} \subset \mathcal{Z}^+$ if $f(n) = 1$ when $n \in \mathcal{S}$ and $f(n) = 0$ when $n \notin \mathcal{S}$. Prove that the characteristic function of the set of positive squarefree integers is μ^2 ($\mu^2(n) = (\mu(n))^2$).

8-3.   Use the methods of this section to find ι_1', the inverse of ι_1.

8-4.* For any complex number s, show that the inverse ι'_s of the corresponding iota function ι_s is the multiplicative function such that $\iota_s(p^\beta) = -p^s$ if $\beta = 1$, and $= 0$ if $\beta > 1$, for every prime p.

8-5. For which of the following sets is the characteristic function of the set a multiplicative function?
   (a) $\{1, p, p^2\}$, prime p
   (b) $\{1, p^2\}$, prime p
   (c) $\{2, 3, 6\}$
   (d) $\{1, 2, 3, 6\}$
   (e) $\{1, 4, 6, 24\}$
   (f) $\{1, 3, 4, 12\}$

8-6. Let f be the characteristic function of a set $\mathscr{S} \subset \mathscr{Z}^+$, $\mathscr{S} \neq \varnothing$. Show that $f \in \mathscr{M}$ if and only if both of the following conditions hold:
   (a) $1 \in \mathscr{S}$;
   (b) if $(a,b) = 1$, $a \in \mathscr{S}$ and $b \in \mathscr{S}$ if and only if $ab \in \mathscr{S}$.

8-7. Define the functions $F_j(n) = (j,n)$ for $j = 0, 1, 2$.
   (a) Which of the functions F_0, F_1, F_2 are in $\mathscr{U}$?
   (b) Which of the functions F_j are in $\mathscr{M}$?
   (c) Are any of the functions F_j iota functions?
   (d) Is any one of F_0, F_1, F_2 the characteristic function of some set $\mathscr{S} \subset \mathscr{Z}^+$?
   (e) Find $F_2 \cdot \mu(3^\beta)$ for every $\beta \in \mathscr{Z}^+$.

8-8. If

$$n = \prod_{i=1}^{k} p_i^{\beta_i},$$

define $\rho(n) = \beta_1 + \cdots + \beta_k$ and $v(n) = k$, $v(1) = \rho(1) = 0$. Let $\lambda(n) = (-1)^{\rho(n)}$ and $\eta(n) = 2^{v(n)}$. Prove

$$\sum_{d|n} \lambda(d)\, \mu(d) = \eta(n), \quad \text{and} \quad \sum_{d|n} \lambda(d)\, \mu\left(\frac{n}{d}\right) = \eta(n)\, \lambda(n).$$

(Hint: η and $\lambda \in \mathscr{M}$; use Theorem 8.1.)

8-9. Use the definitions given in Exercise 8-8 and prove that

$$\sum_{ab=n} (-1)^{\rho(a)}\, 2^{v(b)} = 1;$$

hence prove that $\lambda \cdot \eta = \iota_0$.

9. The Sigma Functions

   Definition. *If s is any complex number, the function $\sigma_s \in \mathscr{A}$ is defined by $\sigma_s = \iota_s \cdot \iota_0$. The functions σ_s are called* sigma functions.

Note. If $s = 1$ we will write σ instead of σ_1; if $s = 0$ we will write τ in place of σ_0. These exceptional notations for the sigma functions corresponding to $s = 0$ and $s = 1$ are introduced so that our notation will agree with established conventions. However, it is usually convenient to discuss the class of sigma functions without treating σ and τ separately.

It is immediate from Theorem 7.1 that every sigma function is multiplicative. Also

$$\sigma_s(n) = \iota_s \cdot \iota_0(n)$$

$$= \sum_{d|n} \iota_s(d)\, \iota_0\!\left(\frac{n}{d}\right)$$

$$= \sum_{d|n} d^s,$$

from which it is clear that $\sigma_s(n)$ denotes the sum of the sth powers of the positive divisors of n. In particular, $\sigma(n)$ is the sum of the divisors of n, and $\tau(n)$ is the number of divisors of n. Furthermore, if p is prime,

$$\sigma_s(p^\beta) = \sum_{j=0}^{\beta} (p^j)^s = \begin{cases} \beta + 1 & \text{if } s = 0, \\[2mm] \dfrac{p^{s(\beta+1)} - 1}{p^s - 1} & \text{if } s \neq 0. \end{cases}$$

Now Theorem 8.1 may be used to show that if

$$1 < n = \prod_{j=1}^{k} p_j^{\beta_j},$$

then

$$\tau(n) = \prod_{j=1}^{k} (\beta_j + 1),$$

$$\sigma(n) = \prod_{j=1}^{k} \frac{p_j^{\beta_j+1} - 1}{p_j - 1},$$

$$\sigma_s(n) = \prod_{j=1}^{k} \frac{p_j^{s(\beta_j+1)} - 1}{p_j^s - 1}, \qquad s \neq 0.$$

It is easy to derive various identities involving these arithmetic functions. For example,

(a) by commutativity we have

$$\sigma_s(n) = \iota_s \cdot \iota_0(n) = \sum_{d|n} \left(\frac{n}{d}\right)^s = n^s \sum_{d|n} \frac{1}{d^s};$$

(b)   by Möbius inversion, $\sigma_s \cdot \mu = \iota_s$ or

$$n^s = \sum_{d \mid n} \sigma_s(d)\, \mu\!\left(\frac{n}{d}\right);$$

(c)   since the inverse of the product of functions is the product of the inverses,

$$\sigma_s' = (\iota_s \cdot \iota_0)' = \iota_s' \cdot \iota_0' = \iota_s' \cdot \mu,$$

and ι_s' was found in Exercise 8-4.

Some classical problems, still not completely solved, have been stated in terms of the σ-function. We will discuss one of these, the problem of *perfect numbers*. Prior to discussing perfect numbers we define a *Mersenne prime* to be any prime q of the form $q = a^b - 1$ where $a, b \in \mathscr{Z}^+$ and $b > 1$. For example, $3 = 2^2 - 1$ and $7 = 2^3 - 1$ are Mersenne primes. By using the identity

$$x^{n+1} - 1 = (x - 1)(x^n + x^{n-1} + \cdots + 1),$$

it is easy to see that if $q = a^b - 1$ is a Mersenne prime, then $a = 2$ and b is prime, because $(a - 1)\mid(a^b - 1)$, and since q is prime, $a - 1 = 1$. Similarly, if b is composite, then b has a divisor d, $1 < d < b$, and $(2^d - 1)\mid q$, contradicting the primality of q. Thus, a Mersenne prime is a prime of the form $2^p - 1$, where p is prime.

A positive integer n is called *perfect* if $\sigma(n) = 2n$. It is not known whether there are any odd perfect integers, but the following theorem gives a complete characterization of all even perfect numbers. The sufficiency was known to Euclid; the necessity was first proved by Euler.

Theorem 9.1. *A necessary and sufficient condition that n be an even perfect number is that $n = 2^{p-1}(2^p - 1)$, where $2^p - 1$ is a Mersenne prime.*

Proof. Suppose $n = 2^{p-1}(2^p - 1)$, with p and $2^p - 1$ prime. Then $\sigma(n) = \sigma(2^{p-1})\,\sigma(2^p - 1) = 2n$.

Conversely, suppose n is even and perfect. Say $n = 2^k m$, $k \geq 1$ and m odd. We have $\sigma(n) = (2^{k+1} - 1)\,\sigma(m) = 2n = 2^{k+1}m$. Then $(2^{k+1} - 1)\mid m$, say $m = (2^{k+1} - 1)M$. Substituting above, we get $\sigma(m) = 2^{k+1}M$. Now m and M are distinct divisors of m, so $\sigma(m) \geq m + M$. But this gives

$$2^{k+1}M = \sigma(m) \geq m + M = 2^{k+1}M,$$

so that m and M are the only divisors of m. Hence $M = 1$ and $m = 2^{k+1} - 1$ is prime; as we saw earlier, $2^{k+1} - 1$ is prime only if $k + 1 = p$ is prime. Thus n is of the form described in the theorem.

EXERCISES
9-1.   Prove that $\tau(n)$ is odd if and only if n is a square.
9-2.   Prove that $\sigma(n)$ is odd if and only if n is a square or two times a square.

9-3.   Prove that for all $n \in \mathscr{Z}^+$

(a) $\displaystyle\sum_{d|n} \tau(d)\,\mu\!\left(\frac{n}{d}\right) = 1,$

(b) $\displaystyle\sum_{d|n} \sigma(d)\,\mu\!\left(\frac{n}{d}\right) = n.$

9-4.   Suppose r and s are complex numbers. Prove that $\sigma_s \cdot \sigma_r = \tau \cdot \iota_s \cdot \iota_r$. Write this identity using the summation notation involved in the definition of the product of functions.

9-5.* Prove that the square of the sum of the first n positive integers is equal to the sum of the cubes of the first n positive integers. If $f \in \mathscr{A}$, define f^k by $f^k(n) = (f(n))^k$. Prove $f \in \mathscr{M}$ implies $f^k \in \mathscr{M}$ for all real k. Combine these results to prove that $(\iota_0 \cdot \iota_0 \cdot \iota_0)^2 = (\tau \cdot \iota_0)^2 = \tau^3 \cdot \iota_0 = (\iota_0 \cdot \iota_0)^3 \cdot \iota_0$.

9-6.   If $m, n \in \mathscr{Z}^+$, then m and n are called *amicable* if $\sigma(m) = \sigma(n) = m + n$. (For example, 220 and 284 are amicable.) Prove that if m is even, n odd, and m and n are amicable, then m is either a square or twice a square, and n is a square.

9-7.   Suppose

$$f(n) = \prod_{d|n} g(d).$$

Take logarithms and apply Möbius inversion to prove that this is equivalent to

$$g(n) = \prod_{d|n} f^{\mu(d)}\!\left(\frac{n}{d}\right).$$

9-8.   Find all solutions of the equation $\tau(n) = 10$, $n \in \mathscr{Z}^+$.

9-9.   Prove that $3|\sigma(3n + 2)$ and $4|\sigma(4n + 3)$ for all $n \geq 0$, $n \in \mathscr{Z}$.

9-10. Prove that $12|\sigma(12n - 1)$ for all $n \in \mathscr{Z}^+$.

9-11. If p is prime, prove $\sigma(p^a)|\sigma(p^b)$ if and only if $(a + 1)|(b + 1)$.

9-12. Prove that

$$\sigma(n) = \sum_{m=1}^{n} \int_0^m \cos\left(\frac{2\pi n[x + 1]}{m}\right) dx.$$

9-13. Prove that $\tau(x) = n$ has infinitely many solutions x for every $n > 1$.

9-14. Show that

$$\prod_{d|n} d = n^{\tau(n)/2}, \qquad n \in \mathscr{Z}^+.$$

9-15. Prove that

$$\prod_{d|n} d = n^2$$

if and only if $n = p^3$ or $n = pq$, where p,q are primes.

10. The Euler φ-Function

Definition. *The function φ, called the Euler φ-function, is defined by* $\varphi = \iota_1 \cdot \mu$.

Since ι_1 and $\mu \in \mathcal{M}$, it follows that $\varphi \in \mathcal{M}$. We proceed to find $\varphi(p^\beta)$ for prime p.

$$\begin{aligned}
\varphi(p^\beta) &= \sum_{d|p^\beta} \iota_1(d)\,\mu\left(\frac{p^\beta}{d}\right) \\
&= \sum_{j=0}^{\beta} \iota_1(p^{\beta-j})\,\mu(p^j) \\
&= \iota_1(p^\beta)\,\mu(1) + \iota_1(p^{\beta-1})\,\mu(p) \\
&= p^\beta - p^{\beta-1} = p^\beta\left(1 - \frac{1}{p}\right)
\end{aligned}$$

Then if

$$1 < n = \prod_{j=1}^{r} p_j^{\beta_j},$$

$$\varphi(n) = \prod_{j=1}^{r} p_j^{\beta_j}\left(1 - \frac{1}{p_j}\right) = n \prod_{j=1}^{r}\left(1 - \frac{1}{p_j}\right).$$

If $\mathscr{S}$ is a finite set, we write $\nu\mathscr{S}$ to denote the number of elements in $\mathscr{S}$. Consider the function γ defined at each $n \in \mathscr{Z}^+$ by

$$\gamma(n) = \nu\{t : 1 \le t \le n, \quad (t,n) = 1\}.$$

We will prove that

$$\sum_{d|n} \gamma\left(\frac{n}{d}\right) = n.$$

Let $n \in \mathscr{Z}^+$ be arbitrary and fixed. We define

$$\mathscr{S} = \{1, 2, \ldots, n\}.$$

Now for each d, $1 \le d \le n$, we define

$$\mathscr{S}_d = \{t : 1 \le t \le n, \quad (t,n) = d\}.$$

We first observe that the collection $\{\mathscr{S}_d : 1 \leq d \leq n\}$ is a *partition* of $\mathscr{S}$, that is a collection of disjoint subsets of $\mathscr{S}$, and every element of $\mathscr{S}$ is in some $\mathscr{S}_d$. If $d \neq e$, $1 \leq d \leq n$ and $1 \leq e \leq n$, suppose $\mathscr{S}_e \cap \mathscr{S}_d \neq \varnothing$. Then there is an integer t, $1 \leq t \leq n$, such that $t \in \mathscr{S}_e \cap \mathscr{S}_d$. But this implies $e = (t,n) = d$, which is impossible. Therefore, if $d \neq e$, then $\mathscr{S}_d \cap \mathscr{S}_e = \varnothing$. Also, if $t \in \mathscr{S}$, let $d = (t,n)$. Since $d|n$ and $1 \leq d \leq n$, we have $t \in \mathscr{S}_d$, so every element of $\mathscr{S}$ is in some one of the subsets.

Since the collection of subsets $\mathscr{S}_d$ is a partition, the number of elements in $\mathscr{S}$ is the sum of the numbers of elements in $\mathscr{S}_d$, $d = 1, \ldots, n$. Notice, however, that if $d \nmid n$, then $\mathscr{S}_d = \varnothing$, because if $t \in \mathscr{S}_d$, then $(t,n) = d$ and $d|n$. Therefore, $\mathbf{v}\mathscr{S}_d = 0$ if $d \nmid n$, and

$$n = \mathbf{v}\mathscr{S} = \sum_{d=1}^{n} \mathbf{v}\mathscr{S}_d = \sum_{d|n} \mathbf{v}\mathscr{S}_d.$$

To complete the proof, we notice that for every d such that $d|n$,

$$\mathbf{v}\mathscr{S}_d = \mathbf{v}\{t : 1 \leq t \leq n, \quad (t,n) = d\}$$

$$= \mathbf{v}\left\{\frac{t}{d} : 1 \leq \frac{t}{d} \leq \frac{n}{d}, \quad \left(\frac{t}{d}, \frac{n}{d}\right) = 1\right\} = \gamma\left(\frac{n}{d}\right).$$

This proves that $\iota_0 \cdot \gamma = \iota_1$. By Möbius inversion, this implies that $\gamma = \iota_1 \cdot \mu$, and we see that $\gamma = \varphi$. We summarize our discussion in Theorem 10.1.

Theorem 10.1. *The function* φ *is a multiplicative function satisfying the identities*

$$\sum_{d|n} \varphi\left(\frac{n}{d}\right) = \sum_{d|n} \varphi(d) = n$$

and

$$\varphi(n) = \sum_{d|n} d\, \mu\left(\frac{n}{d}\right) = n \sum_{d|n} \frac{\mu(d)}{d}.$$

Furthermore, $\varphi(n)$ *is the number of integers* t *such that* $(t,n) = 1$ *and* $1 \leq t \leq n$, *and if*

$$1 < n = \prod_j p_j^{\delta_j}, \qquad \delta_j > 0,$$

then

$$\varphi(n) = n \prod_j \left(1 - \frac{1}{p_j}\right).$$

The Euler φ-function is the best known member of the class of *totient* functions, and is sometimes referred to as the "Euler totient". To define a totient, we first need to point out that a function $f \in \mathscr{A}$ is called *completely multiplicative* if $f(mn) = f(m)f(n)$ for all, $n \in \mathscr{Z}^+$. Clearly, if f is completely multiplicative, then $f \in \mathscr{M}$, but not conversely—for example, $\sigma \in \mathscr{M}$ but $\sigma(4) \neq \sigma(2)\sigma(2)$.

Notice that in Exercise 7-2 it was shown that not only is $\iota_s \in \mathscr{M}$, but in fact ι_s is completely multiplicative for every complex s.

Definition. *A function $f \in \mathscr{A}$ is called a* totient *if and only if $f = g \cdot h'$, where g is completely multiplicative and h' is the inverse of a completely multiplicative function.*

EXERCISES

10-1. Show that φ is a totient.

10-2. If f is a totient, then $f \in \mathscr{M}$.

10-3. Prove the identities

    (a) $\varphi \cdot \tau = \iota_1 \cdot \iota_0$,

    (b) $\varphi \cdot \sigma_s = \iota_1 \cdot \iota_s$,

    (c) $\varphi' = \iota_1' \cdot \iota_0$.

    Write these identities using the summation notation.

10-4. Prove or give a counter example for the following:

    (a)   If f is completely multiplicative, then so is f'.

    (b)   If f and g are completely multiplicative, then so is $f \cdot g$.

    (c)   If f is a totient, then so is f'.

    (d)   If f and g are totients, then so is $f \cdot g$.

10-5. Let $n \in \mathscr{Z}^+$, $k \in \mathscr{Z}$. Prove that the set

$$\{t : k < t \leq k + n, \qquad (t,n) = 1\}$$

    contains exactly $\varphi(n)$ members.

10-6. Find necessary and sufficient conditions that the characteristic function of a set $\mathscr{S} \subset \mathscr{Z}^+$ be completely multiplicative.

10-7. Find all solutions of the equation $\varphi(n) = 10$, $n \in \mathscr{Z}^+$.

10-8. Find $\varphi \cdot \varphi(n)$ for every $n \in \mathscr{Z}^+$. [*Hint:* Since the function $\varphi \cdot \varphi \in \mathscr{M}$, it suffices to find $\varphi \cdot \varphi(p^\beta)$ for every prime p, every $\beta \in \mathscr{Z}^+$.]

10-9. If $k \in \mathscr{Z}^+$, define the arithmetic function J_k to be the product of ι_k and μ, i.e., $J_k = \iota_k \cdot \mu$. The function J_k is called the *Jordan totient*; prove that J_k is in fact a totient for every $k \in \mathscr{Z}^+$. Notice that $J_1 = \varphi$, so that the Jordan totient is a generalization of the Euler totient. Prove that $J_k(1) = 1$, and if

$$1 < n = \prod_j p_j^{\beta_j},$$

then

$$J_k(n) = n^k \prod_j \left(1 - \frac{1}{p_j^k}\right).$$

10-10. If r,s are complex numbers, define $F_{r,s} = \iota_r \cdot \iota_s'$.
  (a) Prove that $F_{r,s}$ is a totient.
  (b) Show that the Jordan totient J_k is a special case of this totient $F_{r,s}$.
  (c) Use the results of Exercise 8-4 to prove that if

$$n = \prod_j p^\beta,$$

$$F_{r,s}(n) = n^r \prod_j \left(1 - \frac{p^s}{p^r}\right).$$

10-11. If p is a prime but $2p + 1$ is not, then show that $\varphi(x) = 2p$ has no solutions.

10-12. If $d|n$ and $k \in \mathscr{Z}^+$, prove that

$$\varphi(nd^k) = d^k\, \varphi(n).$$

10-13. If $(m,n) = d$, show that

$$\varphi(mn)\, \varphi(d) = d\, \varphi(m)\, \varphi(n).$$

10-14. Prove that $\varphi(n)\, \tau(n) \geq n,\ n \in \mathscr{Z}^+$.

Chapter 3

CONGRUENCES AND RESIDUES

11. Complete Residue Systems

If m is a fixed positive integer, the division algorithm, with m as divisor, leads to an interesting classification of all integers into disjoint classes. If $a_1, a_2 \in \mathscr{Z}$, we find the unique q_i, r_i $(i = 1, 2)$ such that

(1)
$$a_1 = q_1 m + r_1, \qquad 0 \le r_1 < m$$
$$a_2 = q_2 m + r_2, \qquad 0 \le r_2 < m.$$

Then we will write $a_1 \equiv a_2 \pmod{m}$, read "a_1 is *congruent* to a_2 modulo m," if and only if $r_1 = r_2$; also $a_1 \not\equiv a_2 \pmod{m}$ means a_1 is not congruent to a_2 modulo m, and r_1 is called the least non-negative *residue* of a_1 modulo m. The first theorem gives a property equivalent to our defining property.

Note. Throughout this chapter we will assume that the modulus (usually to be denoted by m) is larger than 1.

Theorem 11.1. *A necessary and sufficient condition that $a_1 \equiv a_2 \pmod{m}$ is that $m | (a_1 - a_2)$.*

Proof. Suppose that the division algorithm gives the system of equations (1) and suppose $a_1 \equiv a_2 \pmod{m}$. Then $r_1 = r_2$ implies that $a_1 - q_1 m = a_2 - q_2 m$, and $m | (a_1 - a_2)$. Conversely, if $m | (a_1 - a_2)$, then for some b,

$$bm = a_1 - a_2 = (q_1 - q_2)m + r_1 - r_2,$$

from which it follows that $m | (r_1 - r_2)$. But from the inequalities in (1) we have $-m < r_1 - r_2 < m$, and the only multiple of m in this range is $0 = r_1 - r_2$; thus the proof is complete.

The relation "congruence" on the set $\mathscr{Z}$ of integers is an example of an *equivalence relation* on a set. In general, the relation $\mathscr{R}$ on a set $\mathscr{S}$ is an

31

equivalence relation if

(a) $\mathscr{R}$ is reflexive, i.e., $a\mathscr{R}a$ for all $a \in \mathscr{S}$;

(b) $\mathscr{R}$ is symmetric, i.e., if $a\mathscr{R}b$, then $b\mathscr{R}a$;

(c) $\mathscr{R}$ is transitive, i.e., if $a\mathscr{R}b$ and $b\mathscr{R}c$, then $a\mathscr{R}c$.

It is well known that if $\mathscr{S}$ is a set and $\mathscr{R}$ is a given equivalence relation on $\mathscr{S}$, then $\mathscr{R}$ induces a partition of $\mathscr{S}$, and that, conversely, if $\mathscr{F}$ is any partition of $\mathscr{S}$, then $\mathscr{F}$ may be used to define an equivalence relation on $\mathscr{S}$. More precisely, we have the following.

Theorem 11.2. *Suppose $\mathscr{S}$ is a set and $\mathscr{R}$ is an equivalence relation on $\mathscr{S}$. For every $a \in \mathscr{S}$, the subset $\mathscr{S}_a$ is defined by*

$$\mathscr{S}_a = \{b \in \mathscr{S} : b\mathscr{R}a\}.$$

Then the collection $\mathscr{F} = \{\mathscr{S}_a : a \in \mathscr{S}\}$ is a partition of $\mathscr{S}$. Conversely, if $\mathscr{F}$ is any partition of a set $\mathscr{S}$, then the relation $\mathscr{R}$ defined by "if $a, b \in \mathscr{S}$, then $a\mathscr{R}b$ if and only if there exists a set $\mathscr{T} \in \mathscr{F}$ such that $a \in \mathscr{T}$ and $b \in \mathscr{T}$" is an equivalence relation on $\mathscr{S}$.

Proof. Suppose $\mathscr{R}$ is an equivalence relation on $\mathscr{S}$. Evidently

$$\bigcup_{\mathscr{S}_a \in \mathscr{F}} \mathscr{S}_a = \bigcup_{a \in \mathscr{S}} \mathscr{S}_a = \mathscr{S}.$$

so every element of $\mathscr{S}$ is in some subset $\mathscr{S}_a$. Assume that $\mathscr{S}_a, \mathscr{S}_b \in \mathscr{F}$ and $\mathscr{S}_a \cap \mathscr{S}_b \neq \varnothing$. Then there is some $c \in \mathscr{S}$ such that $c \in \mathscr{S}_a \cap \mathscr{S}_b$, and $c \in \mathscr{S}_a$ implies $c\mathscr{R}a$ and $c \in \mathscr{S}_b$ implies $c\mathscr{R}b$. Therefore, $a\mathscr{R}c$ by symmetry; together with $c\mathscr{R}b$, transitivity gives $a\mathscr{R}b$. Again by symmetry, we also have $b\mathscr{R}a$.

Let $x \in \mathscr{S}_a$. Then $x\mathscr{R}a$ and $a\mathscr{R}b$ imply $x\mathscr{R}b$, so $x \in \mathscr{S}_b$ and $\mathscr{S}_a \subset \mathscr{S}_b$. Similarly, if $y \in \mathscr{S}_b$, then $y\mathscr{R}b$ and $b\mathscr{R}a$ imply $y \in \mathscr{S}_a$, so $\mathscr{S}_b \subset \mathscr{S}_a$, hence $\mathscr{S}_a = \mathscr{S}_b$. Contrapositively, any two distinct subsets are disjoint. Therefore, $\mathscr{F}$ is a partition.

Conversely, assume $\mathscr{F}$ is any partition of $\mathscr{S}$ and consider $\mathscr{R}$ defined as in the statement of the theorem. For every $a \in \mathscr{S}$, a is in some subset, and therefore $a\mathscr{R}a$. Clearly $a\mathscr{R}b$ implies a and b are in the same subset, so $b\mathscr{R}a$. Finally, if $a\mathscr{R}b$ and $b\mathscr{R}c$, then a, b, c are in the same subset, so $a\mathscr{R}c$, and $\mathscr{R}$ is an equivalence relation. This completes the proof of the theorem.

The subsets $\mathscr{S}_a$ are called *equivalence classes*, and any element in $\mathscr{S}_a$ is called a representative of the class. When an operation, say $\times$, has been defined on $\mathscr{S}$, then it may be possible to define an analogous operation on the set $\mathscr{F}$ of equivalence classes: if $a\mathscr{R}b$ and $c\mathscr{R}d$ imply that $(a \times c)\mathscr{R}(b \times d)$, then we can define

$$\mathscr{S}_a \times \mathscr{S}_c = \mathscr{S}_{a \times c}.$$

In Theorem 11.3 we will show in effect that it is possible to define $+$ and $\cdot$ on the equivalence classes $\mathscr{Z}_a$ induced by the congruence relation.

In the case of the congruence (mod m) relation on $\mathscr{Z}$, we agree to denote an equivalence class $\mathscr{Z}_a$ by *any* one of its representatives; thus the operations on the equivalence classes (also called *residue classes*) are defined so that the sum (or product) of the class containing $a \in \mathscr{Z}$ and the class containing $b \in \mathscr{Z}$ is the class containing $a + b$ (or ab). The resulting system is denoted $\mathscr{Z}/(m) = (\mathscr{Z}/(m), +, \cdot)$, and is called the system of *integers* (mod m).

For example, take $m = 5$. The residue classes are

$$\mathscr{Z}_0 = \{\ldots, -10, -5, 0, 5, 10, \ldots\} = Z_{-5} = Z_5 = \ldots$$

$$\mathscr{Z}_1 = \{\ldots, -9, -4, 1, 6, 11, \ldots\}$$

$$\mathscr{Z}_2 = \{\ldots, -8, -3, 2, 7, 12, \ldots\}$$

$$\mathscr{Z}_3 = \{\ldots, -7, -2, 3, 8, 13, \ldots\}$$

$$\mathscr{Z}_4 = \{\ldots, -6, -1, 4, 9, 14, \ldots\}.$$

Notice that $\mathscr{Z} = \mathscr{Z}_0 \cup \mathscr{Z}_1 \cup \cdots \cup \mathscr{Z}_4$ and if $i \neq j$ $(0 \leq i, j \leq 4)$, then $\mathscr{Z}_i \cap \mathscr{Z}_j = \varnothing$. We denote the class $\mathscr{Z}_i$ by any of its representatives; for instance, -24 or -4 or 6 or 26 may be used to denote $\mathscr{Z}_1$. Instead of writing $\mathscr{Z}_2 + \mathscr{Z}_3 = \mathscr{Z}_{2+3} = \mathscr{Z}_5 = \mathscr{Z}_0$, we write $2 + 3 \equiv 5 \equiv 0 \,(\text{mod } 5)$, and instead of $\mathscr{Z}_2 \cdot \mathscr{Z}_3 = \mathscr{Z}_6 = \mathscr{Z}_1$, we write $2 \cdot 3 \equiv 6 \equiv 1 \,(\text{mod } 5)$. The algebraic system $\mathscr{Z}/(5)$ is the set with five elements, say $\mathscr{Z}_0, \mathscr{Z}_1, \mathscr{Z}_2, \mathscr{Z}_3, \mathscr{Z}_4$ together with the operations of addition and multiplication.

Theorem 11.3. *If* $a \equiv b \,(\text{mod } m)$ *and* $c \equiv d \,(\text{mod } m)$, *then* $a \pm c \equiv b \pm d$ *(mod m) and* $ac \equiv bd \,(\text{mod } m)$.

Proof. If $m|(a - b)$ and $m|(c - d)$, then $m|\{a \pm c - (b \pm d)\}$. To prove that $ac \equiv bd$, say $mA = a - b$ and $mD = c - d$. Then

$$m^2 AD = ac - bd - b(c - d) - d(a - b),$$

from which it is clear that $m|(ac - bd)$.

Corollary 11.3a. *If* $a \equiv b \,(\text{mod } m)$, *then for every* $k \in \mathscr{Z}$, $ak \equiv bk \,(\text{mod } m)$.

Corollary 11.3b. *If* $a \equiv b \,(\text{mod } m)$, *then for every* $k \in \mathscr{Z}^+$, $a^k \equiv b^k \,(\text{mod } m)$.

Theorem 11.4. *Suppose* $(k,m) = d$. *Then* $ka \equiv kb \,(\text{mod } m)$ *if and only if* $a \equiv b \,(\text{mod } m/d)$.

Proof. If $(k,m) = d$, say $dM = m$ and $dK = k$. If $m|(ka - kb)$, we have $mA = k(a - b)$ for some A. It follows that $MA = K(a - b)$ and since $(M, K) = 1$, $M|(a - b)$. But $M = m/d$, so $a \equiv b \,(\text{mod } m/d)$. Conversely, if $a \equiv b \,(\text{mod } M)$, then $M|(a - b)$, so $MB = a - b$ for some B. We have

$$ka - kb = kMB = KdMB = mKB,$$

and $m|(ka - kb)$.

If a is any integer and $m > 1$ is given, it is clear from the division algorithm that there are at most m possible values for the remainder r, and hence at most m residue classes in $\mathscr{Z}/(m)$. On the other hand, there are at least m classes because no two of the numbers $0, 1, \ldots, m - 1$ are congruent (mod m), so there are always exactly m distinct residue classes. A set of m integers which contains exactly one representative of each class is called a *complete residue system modulo m*, abbreviated CRS (mod m). The set $\{1, 2, \ldots, m\}$ is called the *least positive* CRS (mod m), the set $\{0, 1, \ldots, m - 1\}$ is the *least non-negative* CRS (mod m), and the set

$$\left\{ \left[\frac{-m + 2}{2} \right], \left[\frac{-m + 2}{2} \right] + 1, \ldots, 0, 1, \ldots, \left[\frac{m}{2} \right] \right\}$$

is the *absolutely least* CRS (mod m).

For example, if $m = 6$, the absolutely least CRS is the set $\{-2, -1, 0, 1, 2, 3\}$. If $m = 7$, it is $\{-3, -2, \ldots, 2, 3\}$.

Theorem 11.5. *If* $\{a_1, \ldots, a_m\}$ *is a CRS* (mod m), *and if* $(k,m) = 1$, *then* $\{ka_1, \ldots, ka_m\}$ *is also a CRS* (mod m).

Proof. Suppose that for some $i \neq j$ we have

$$ka_i \equiv ka_j \,(\text{mod } m).$$

Since $(k,m) = 1$, Theorem 11.4 implies $a_i \equiv a_j \,(\text{mod } m)$. But this contradicts the fact that the numbers $a_1, \ldots, a_m$ form a CRS. Hence the set $\{ka_1, \ldots, ka_m\}$ is a set of m pairwise incongruent elements, and so is a CRS (mod m).

EXERCISES

11-1.   Prove that congruence modulo m is an equivalence relation on $\mathscr{Z}$.

11-2.   Prove Corollary 11.3a and Corollary 11.3b.

11-3.*   If $a \equiv b \,(\text{mod } m)$ and $0 < d|m$, prove $a \equiv b \,(\text{mod } d)$.

11-4.   Find x in the least non-negative CRS (mod 7) such that

(a) $8^{25} \equiv x \,(\text{mod } 7)$,

(b) $2^{30} \equiv x \,(\text{mod } 7)$,

(c) $(-4)^{10} \equiv x \,(\text{mod } 7)$.

11-5.   Suppose $\mathscr{S} = (\mathscr{S}, +, \cdot)$ is an algebraic system and let $\mathscr{S}'$ denote the subset of $\mathscr{S}$ obtained by deleting from $\mathscr{S}$ the identity with respect to the operation "$+$". Then $\mathscr{S}$ is said to be a *commutative ring with identity* if $(\mathscr{S}, +)$ is a commutative group, $(\mathscr{S}', \cdot)$ is a commutative semigroup with identity, and "$\cdot$" is distributive over "$+$", that is, for all $a, b, c \in \mathscr{S}$, $a \cdot (b + c) = a \cdot b + a \cdot c$. $\mathscr{S}$ is a *field* if both $(\mathscr{S}, +)$ and $(\mathscr{S}', \cdot)$ are commutative groups and the distributive law holds.

(a)   Prove $(\mathscr{Z}/(m), +, \cdot)$ is a commutative ring with identity.

(b)   If m is prime, prove $\mathscr{Z}/(m)$ is a field.

(c)   Prove that if $\mathscr{L}/(m)$ is a field, then m is prime.

11-6.   Show that the set

$$\mathscr{S} = \left\{ \left[\frac{-m+2}{2}\right], \left[\frac{-m+2}{2}\right] + 1, \ldots, 0, 1, \ldots, \left[\frac{m}{2}\right] \right\}$$

contains exactly m distinct elements, no two of which are congruent modulo m. Thus prove that the absolutely least CRS $(\bmod\, m)$ is a CRS $(\bmod\, m)$.

11-7.   Give an example to show that for some a, b, k, m we may have $ka \equiv kb\,(\bmod\, m)$, but $a \not\equiv b\,(\bmod\, m)$.

11-8.   Give an example to show that Theorem 11.5 may not be true if $(k,m) \neq 1$. Prove or disprove the following statement: If $\{a_1, \ldots, a_m\}$ is a CRS $(\bmod\, m)$, and if $(k,m) > 1$, then $\{ka_1, \ldots, ka_m\}$ is *not* a CRS $(\bmod\, m)$.

11-9.*   If $b \equiv c\,(\bmod\, m_1)$ and $b \equiv c\,(\bmod\, m_2)$, prove $b \equiv c\,(\bmod\, \langle m_1, m_2 \rangle)$.

12. Linear Congruences

We will find necessary and sufficient conditions for the existence of a solution x of the congruence $ax + c \equiv 0\,(\bmod\, m)$, and show that this is related to the problem of finding solutions x,y of the linear Diophantine equation $ax + by = n$. (By a Diophantine equation we shall mean an equation, in one or more unknowns, whose solutions are required to be integers.) Later in this section we will show how to solve a system of linear congruences.

If $f(x)$ is a polynomial with integral coefficients, we say x_0 is a solution of the congruence $f(x) \equiv 0\,(\bmod\, m)$ if $f(x_0) \equiv 0\,(\bmod\, m)$. The problem of finding solutions of a congruence is very different from finding solutions of polynomial equations. For example, the equation $x^3 - 3x^2 + 2x = 0$ has three solutions, $x = 0, 1, 2$, but the congruence

$$x^3 - 3x^2 + 2x \equiv 0\,(\bmod\, 3)$$

is satisfied by every integer. As another example, $x - 2 = 0$ has a unique solution, but $x - 2 \equiv 0\,(\bmod\, 5)$ is solved by every $x \equiv 2\,(\bmod\, 5)$. Finally, $x^2 - 2 = 0$ has two roots, but the congruence $x^2 - 2 \equiv 0\,(\bmod\, 3)$ has no solution. These examples show that a polynomial congruence may have no solutions or infinitely many solutions. That there are no other possibilities is implied by our next theorem.

Theorem 12.1. *Suppose $f(x)$ is a polynomial with integral coefficients. If $a \equiv b\,(\bmod\, m)$, then $f(a) \equiv f(b)\,(\bmod\, m)$.*

Proof. Let

$$f(x) = \sum_{k=0}^{n} c_k x^k.$$

From Corollary 11.3b, we know that $a \equiv b \pmod{m}$ implies $a^k \equiv b^k \pmod{m}$ for $k = 1, \ldots, n$, and from Corollary 11.3a, that $c_k a^k \equiv c_k b^k$ for $k = 1, \ldots, n$. Therefore,

$$\sum_{k=0}^{n} c_k a^k \equiv \sum_{k=0}^{n} c_k b^k \pmod{m};$$

but this says $f(a) \equiv f(b) \pmod{m}$.

Corollary 12.1. *If a is a solution of $f(x) \equiv 0 \pmod{m}$, then for every b such that $b \equiv a \pmod{m}$, b is also a solution.*

Because of this last result, we agree to say *a congruence has n solutions* if it has exactly n *incongruent* solutions. If m is small, solutions can be found by testing each number in a CRS $\pmod{m}$, but for large m this method is impractical. For a linear congruence one may always use the Euclidean algorithm to find a solution (if one exists), and we will show that it is possible to get all solutions from a given one. Moreover, we can easily tell how many solutions will exist. The problem of finding solutions of a congruence is somewhat easier if we first prove this result on solutions of a linear Diophantine equation.

Theorem 12.2. *The equation $ax + by = n$ has a solution x,y if and only if $d = (a,b)|n$. If there is a solution x_0, y_0, then there are infinitely many solutions: for every $t \in \mathscr{Z}$, then x,y of the form $x = x_0 + (b/d)_t, y = y_0 + (-a/d)t$ is a solution, and every solution x,y is of this form for some $t \in \mathscr{Z}$.*

Proof. Since the first part of this theorem is merely a restatement of Corollary 2.1a, we proceed with the proof of the second part. Suppose that x_0, y_0 is a solution. For every $t \in \mathscr{Z}$,

$$a\left(x_0 + \frac{b}{d}t\right) + b\left(y_0 - \frac{a}{d}t\right) = ax_0 + by_0 = n,$$

so $x_0 + (b/d)_t, y_0 + (a/d)(-t)$ *is a solution.* If x,y is any solution, then we define X and Y by $X = x - x_0$, $Y = y - y_0$. Then

$$n = ax + by = a(X + x_0) + b(Y + y_0)$$

$$= ax_0 + by_0 + aX + bY.$$

This gives $aX + bY = 0$, from which we have

$$(2) \qquad \frac{a}{d}X + \frac{b}{d}Y = 0.$$

Therefore, $(b/d)|(a/d)X$, but $(a/d, b/d) = 1$, so $(b/d)|X$; say $(b/d)t = X$. Then (2) implies $Y = -(a/d)t$. Thus $x = x_0 + X = x_0 + (b/d)t$ and $y = y_0 - (a/d)t$ as required.

Corollary 12.2. *The linear congruence* $ax \equiv n \pmod{m}$ *has solutions if and only if* $d = (a,m)|n$. *If there is a solution, then there are exactly* d *(incongruent) solutions. In particular, if* $(a,m) = 1$, *then the congruence has a unique solution.*

Proof. The first part is evident since there exists an x such that $ax \equiv n \pmod{m}$ if and only if there exist x and y such that $ax + my = n$. If there is a solution x_0 of the congruence, then all solutions x are of the form $x = x_0 + (m/d)t$, $t \in \mathscr{Z}$. If

$$x_1 = x_0 + \frac{m}{d}t_1 \quad \text{and} \quad x_2 = x_0 + \frac{m}{d}t_2$$

are solutions, and if $x_1 \equiv x_2 \pmod{m}$, then

$$x_0 + \frac{m}{d}t_1 \equiv x_0 + \frac{m}{d}t_2 \pmod{m}$$

$$t_1 \equiv t_2 \left(\text{mod}\, \frac{m}{(m,m/d)}\right)$$

$$t_1 \equiv t_2 \pmod{d}.$$

Therefore, two solutions, x_1 and x_2, are incongruent $\pmod{m}$ if and only if t_1 and t_2 are incongruent $\pmod{d}$, so there are d distinct ways to obtain incongruent solutions.

Example. Consider the congruence $3x \equiv 9 \pmod{12}$. By Theorem 11.4, this is equivalent to $x \equiv 3 \pmod{4}$, which obviously has the solution $x = 3$. Then the original congruence will have $(3,12) = 3$ solutions, all of the form $3 + (12/3)t$. Therefore, $x = 3,7,11$ are the solutions in the least positive CRS $\pmod{12}$ and $x = -5,-1,3$ are the solutions in the absolutely least CRS.

The following result gives a method for finding a solution of a system of linear congruences under certain conditions.

Theorem 12.3. (*Chinese Remainder Theorem.*) *If* $(m_i, m_j) = 1$ *for* $1 \leq i < j \leq r$, *then the system*

(3)
$$\begin{cases} x \equiv a_1 \pmod{m_1} \\ x \equiv a_2 \pmod{m_2} \\ \quad \cdots \\ x \equiv a_r \pmod{m_r} \end{cases}$$

always has a solution x which is unique modulo

$$\prod_{i=1}^{r} m_i.$$

Proof. Let

$$M = \prod_{i=1}^{r} m_i$$

and $M = A_i m_i$, $i = 1, \ldots, r$. For each i, $(A_i, m_i) = 1$. Let y_i be the unique (mod m_i) solution of

$$A_i y_i \equiv 1 \,(\text{mod } m_i).$$

Notice that if $i \neq j$, $A_j \equiv 0 \,(\text{mod } m_i)$, so that

$$x = \sum_{i=1}^{r} a_i A_i y_i$$

is a solution of (3).

Suppose that x and z are any two solutions of (3). For every i, $x \equiv a_i \equiv z$ (mod m_i) implies that $m_i | (x - z)$ and hence, $M | (x - z)$ by Exercise 2-1(b). That is, any two solutions are congruent (mod M).

Example. Solve

(4)
$$\begin{cases} 2x \equiv 1 \,(\text{mod } 3) \\ 3x \equiv 2 \,(\text{mod } 4) \\ 4x \equiv 2 \,(\text{mod } 5). \end{cases}$$

Evidently we must have

(5)
$$\begin{cases} x \equiv 2 \,(\text{mod } 3) \\ x \equiv 2 \,(\text{mod } 4) \\ x \equiv 3 \,(\text{mod } 5). \end{cases}$$

Now we use the Chinese Remainder Theorem to construct the solution of (5). We find $M = 60$, $A_1 = 20$, $A_2 = 15$, $A_3 = 12$, $y_1 = 2$, $y_2 = y_3 = 3$. Then $\Sigma a_i A_i y_i = 278 \equiv 38 \,(\text{mod } 60)$. Thus, $x = 38$ is the solution of both (4) and (5).

EXERCISES

12-1. Find all solutions x, y of $2x + 3y = 7$ such that $-10 < x < 10$.

12-2. For the congruence $12x \equiv 16 \,(\text{mod } 28)$, find all solutions in

    (a) the least positive CRS (mod 28),

    (b) the absolutely least CRS (mod 28).

12-3. Solve the system $x \equiv 1 \pmod 3$, $x \equiv 2 \pmod 5$, $x \equiv 3 \pmod 7$.

12-4. Solve the system $3x \equiv 1 \pmod 4$, $2x \equiv 4 \pmod 6$, $2x \equiv 3 \pmod{11}$.

12-5. Sketch the graph of the equation $4x + 6y = 10$. On this line, call a point P a "solution point" if P has abscissa x, ordinate y, and x,y is a solution of the Diophantine equation $4x + 6y = 10$. Find a solution point which is

(a) closer to the origin than every other solution point;

(b) closer to the y–axis than every other solution point;

(c) at least as close to the x–axis as every other solution point.

12-6. Consider the graph of $ax + by = n$, where a, b, $n \in \mathscr{Z}$. Prove that the set of solution points (see Exercise 12-5) is just the set of points on the line $ax + by = n$ which have integer coordinates.

13. Reduced Residue Systems

If $\mathscr{S}$ is any CRS $\pmod m$, the subset of $\mathscr{S}$ which consists of those $a \in \mathscr{S}$ such that $(a,m) = 1$ is called a *reduced residue system modulo m*, abbreviated RRS $\pmod m$. As with a CRS, we may speak about the least positive or the absolutely least RRS $\pmod m$. Evidently the set $\{a : 1 \le a \le m, (a,m) = 1\}$ is the least positive RRS $\pmod m$, and the set contains exactly $\varphi(m)$ elements. If $\mathscr{R}$ is any other RRS $\pmod m$, then $\mathscr{R}$ also contains $\varphi(m)$ elements and every element of $\mathscr{R}$ is congruent $\pmod m$ to exactly one element in the least positive RRS. We can see this is so as follows. Suppose $b \in \mathscr{R}$, and consider the congruence $x \equiv b \pmod m$. This has a unique solution x_0 in the least positive CRS $\pmod m$. But since $x_0 - b = km$ for some k, if $(x_0,m) = d > 1$, then $d|(m,b)$, which contradicts that b is in a RRS $\pmod m$. Therefore, $(x_0,m) = 1$ and x_0 is in the least positive RRS.

Lemma 13.1. *If $\{a_1, \dots, a_{\varphi(m)}\}$ is a RRS $(mod\ m)$ and if $(k,m) = 1$, then $\{ka_1, \dots, ka_{\varphi(m)}\}$ also is a RRS $(mod\ m)$.*

The proof of the lemma is the same as the proof of Theorem 11.5. The lemma leads to this important result.

Theorem 13.1. (*Euler's Theorem.*) *If $(k,m) = 1$, then*

$$k^{\varphi(m)} \equiv 1 \pmod m.$$

Proof. Let $\{a_1, \dots, a_{\varphi(m)}\}$ be a RRS $\pmod m$. Then also $\{ka_1, \dots, ka_{\varphi(m)}\}$ is a RRS $\pmod m$, so that in some order the a_i are congruent $\pmod m$ to the ka_j. Therefore,

$$\prod_{i=1}^{\varphi(m)} (ka_i) \equiv \prod_{i=1}^{\varphi(m)} a_i \pmod m;$$

$$k^{\varphi(m)} \prod a_i \equiv \prod a_i \pmod m.$$

But $(a_i,m) = 1$ for $i = 1,\dots,\varphi(m)$ and $(\Pi a_i,m) = 1$ so we may cancel this common factor from the last congruence without affecting the modulus.

A special case of Euler's Theorem was first proved by Fermat.

Corollary 13.1. (*Fermat's Theorem.*) *If p is a prime, then*

$$a^p \equiv a \,(\text{mod } p).$$

The Fermat (as well as the Euler) Theorem may be used to simplify computations. For example, if asked to show that $23^{475} \equiv 2 \,(\text{mod } 7)$, we would notice that $23^{475} \equiv 2^{475} \equiv 2^{6(79)+1} \equiv (2^6)^{79} \cdot 2 \equiv 1^{79} \cdot 2 \equiv 2 \,(\text{mod } 7)$.

Suppose now that for some $k, k^{\varphi(m)} \equiv 1 \,(\text{mod } m)$. Then the congruence $kx \equiv 1 \,(\text{mod } m)$ has a solution $x = k^{\varphi(m)-1}$, so by Corollary 12.2, $(k,m)|1$, or k and m are relatively prime. This shows that the converse of the Euler Theorem is true, namely, if $k^{\varphi(m)} \equiv 1 \,(\text{mod } m)$, then $(k,m) = 1$.

Our next theorem will be applied in Section 14 in the study of another arithmetic function.

Theorem 13.2. *Suppose $(m,n) = 1$. Let $\mathscr{R}_1$ be a RRS (mod m) and $\mathscr{R}_2$ be a RRS (mod n). Then the set*

$$\mathscr{R} = \{km + jn : k \in \mathscr{R}_2, j \in \mathscr{R}_1\}$$

is a RRS (mod mn).

Proof. There are $\mathbf{v}\mathscr{R}_2 = \varphi(n)$ choices for k and $\mathbf{v}\mathscr{R}_1 = \varphi(m)$ choices for j, so at most $\varphi(m)\,\varphi(n) = \varphi(mn)$ elements in $\mathscr{R}$. That there are exactly $\varphi(mn)$ elements may be shown by an indirect argument. Suppose for some $j_1, j_2 \in \mathscr{R}_1$ and some $k_1, k_2 \in \mathscr{R}_2, k_1 \leq k_2$, we have

$$k_1 m + j_1 n = k_2 m + j_2 n.$$

Then

$$(6) \qquad\qquad (j_1 - j_2)n = (k_2 - k_1)m,$$

so $n|(k_2 - k_1)$. But since the k_i are from a RRS (mod n), this implies $k_1 = k_2$. Then (6) yields $j_1 = j_2$. Thus, $\mathbf{v}\mathscr{R} = \varphi(mn)$.

We now show that no two elements of $\mathscr{R}$ are congruent (mod mn). Suppose instead that for $k_i, j_i \,(i = 1,2)$ as above, we have

$$k_1 m + j_1 n - (k_2 m + j_2 n) = qmn$$

for some q. Then $(k_1 - k_2)m = (qm - j_1 + j_2)n$. But, as above, $n|(k_1 - k_2)$ implies $k_1 = k_2$ and it follows that $j_1 = j_2$. This shows that $\mathscr{R}$ is contained in some CRS (mod mn).

Finally, assume $km + jn \in \mathscr{R}$, and say $d = (km + jn, mn)$. Now d divides any linear combination of these numbers, and since $km^2 = m(km + jn) - j(mn), d|km^2$. Write $d = ab$, where $a|m, b|n$. Since $b|d$, also $b|km^2$. But $(b,m) = 1$, so $b|k$, and $b|n$, therefore $b|(k,n) = 1$. In a similar manner, $a = 1$, so $d = 1$. Thus, $\mathscr{R}$ consists of $\varphi(mn)$ elements from a CRS (mod mn), each of which is relatively prime to mn, so $\mathscr{R}$ is a RRS (mod mn).

EXERCISES

13-1. Give a proof of Fermat's Theorem.

13-2. In $\mathscr{Z}/(m)$ consider the subset of elements $\mathscr{Z}_i$ with $(i,m) = 1$. Prove that this subset is a group under multiplication. [This result may be stated roughly by saying that a RRS (mod m) is a group under multiplication if one recognizes that the product of two elements is equivalent to any congruent integer.]

13-3.* Suppose for some $k \in \mathscr{Z}$, $n \in \mathscr{Z}^+$ we have

(7) $$k^n \equiv 1 \ (\text{mod } m).$$

Prove that $(k,m) = 1$. If n is the smallest such integer for which (7) holds, prove that $n|\varphi(m)$. [*Hint:* Use the division algorithm on n and $\varphi(m)$.]

13-4. Suppose $(m,n) = 1$. Let $\mathscr{C}_1$ be a CRS (mod m) and $\mathscr{C}_2$ be a CRS (mod n). Prove that the set

$$\mathscr{C} = \{km + jn : k \in \mathscr{C}_2, j \in \mathscr{C}_1\}$$

is a CRS (mod mn).

13-5. If $(a,m) = 1$, show that the solution of $ax \equiv n$ (mod m) is

$$x = a^{\varphi(m)-1}n.$$

13-6. If $\{a_1, \ldots, a_{\varphi(m)}\}$ is a RRS (mod m) and if $(k,m) > 1$, prove that $\mathscr{K} = \{ka_1, \ldots, ka_{\varphi(m)}\}$ is *not* a RRS (mod m). Does there exist a non-empty subset of $\mathscr{K}$ which is contained in some RRS (mod m)?

13-7. Suppose $\{a_1, \ldots, a_{\varphi(m)}\}$ is a RRS (mod m). Use the division algorithm to find q_j, r_j [$j = 1, \ldots, \varphi(m)$] such that $a_j = mq_j + r_j$, $0 \leq r_j < m$. Prove that $\{r_1, \ldots, r_{\varphi(m)}\}$ is the least positive RRS (mod m).

13-8. Using $\{1,3\}$ and $\{1,2,3,4\}$ for the RRS (mod 4) and (mod 5), respectively, compute the corresponding RRS (mod 20) obtained by the method of Theorem 13.2. Notice that though we started with the least positive RRS (mod 4) and the least positive RRS (mod 5), we do not get the least positive RRS (mod 20).

14. Ramanujan's Trigonometric Sum

Trigonometric functions are encountered frequently in number theory. As a typical example we consider Ramanujan's trigonometric sum. We assume the reader is familiar with the following properties of the exponential function of a complex variable:

(8) $\qquad\qquad e^{z_1 + z_2} = e^{z_1}e^{z_2}$ and $(e^{z_1})^{z_2} = e^{z_1 z_2};$

(9) $\qquad\qquad$ if z is real, $e^{iz} = \cos z + i \sin z.$

It will be convenient here to let $\mathcal{R}_n$ denote the least positive RRS (mod n).

Definition. *If* $s \in \mathscr{L}^+$, *the function* $c_s \in \mathscr{A}$ *is defined by*

$$c_s(n) = \sum_{k \in \mathcal{R}_n} e^{2\pi i s k/n};$$

the function c_s *is called* Ramanujan's trigonometric sum.

We will first show that if the numbers k in the summation are from *any* RRS (mod n), the value $c_s(n)$ is not changed.

Lemma 14.1. *If* $\mathcal{T}$ *is any RRS* (*mod n*), *then*

$$c_s(n) = \sum_{t \in \mathcal{T}} e^{2\pi i s t/n}.$$

Proof. We know that if $\mathcal{T}$ is any RRS (mod n) and $\mathcal{R}_n$ is the least positive RRS (mod n), then for each $t \in \mathcal{T}$, there is a unique $k \in \mathcal{R}_n$ such that $t \equiv k$ (mod n). Thus, for every $t \in \mathcal{T}$, there is a unique k such that

$$t = k + n\, q(k) \text{ for some integer } q(k).$$

Therefore,

$$\sum_{t \in \mathcal{T}} e^{2\pi i s t/n} = \sum_{k \in \mathcal{R}_n} e^{2\pi i s(k + nq(k))/n}$$

$$= \sum_{k \in \mathcal{R}_n} e^{2\pi i s k/n} e^{2\pi i s q(k)}$$

$$= c_s(n)$$

because, by (9), for every k, $e^{2\pi i s q(k)} = 1$.

We can now prove the following theorem which makes it clear that c_s is a real-valued function.

Theorem 14.1.

$$c_s(n) = \sum_{k \in \mathcal{R}_n} \cos \frac{2\pi s k}{n}.$$

Proof. Take $\mathcal{T} = \{-k : k \in R_n\}$. [That $\mathcal{T}$ is a RRS (mod n) is left as an exercise.] By Lemma 14.1,

$$\sum_{k \in \mathcal{R}_n} e^{2\pi i s k/n} = \sum_{k \in \mathcal{R}_n} e^{2\pi i s(-k)/n}$$

$$\sum_k \left\{ \cos \frac{2\pi s k}{n} + i \sin \frac{2\pi s k}{n} \right\} = \sum_k \left\{ \cos \frac{2\pi s(-k)}{n} + i \sin \frac{2\pi s(-k)}{n} \right\}$$

$$\sum_k \cos \frac{2\pi sk}{n} + i \sum_k \sin \frac{2\pi sk}{n} = \sum_k \cos \frac{2\pi sk}{n} - i \sum_k \sin \frac{2\pi sk}{n}.$$

In the last equation we have used the familiar facts that $\cos(-x) = \cos x$ and $\sin(-x) = -\sin x$ for all real x. From the equation above it is clear that

$$(10) \qquad \sum_{k \in \mathcal{R}_n} \sin \frac{2\pi sk}{n} = 0.$$

Applying (9) to the definition of $c_s(n)$ and using (10), the theorem is proved.

Corollary 14.1. *If $n|s$, then $c_s(n) = \varphi(n)$.*
Proof. Say $nd = s$. Then

$$c_s(n) = \sum_{k \in \mathcal{R}_n} \cos 2\pi \, dk = \sum_{k \in \mathcal{R}_n} 1 = \varphi(n).$$

Before giving an evaluation of $c_s(n)$ it is helpful to prove that c_s is multiplicative.

Theorem 14.2. *For every s, $c_s \in \mathcal{M}$.*
Proof. That $c_s(1) = 1$ follows from the Corollary 14.1. Assume $(m,n) = 1$. Then

$$c_s(m)c_s(n) = \sum_{j \in \mathcal{R}_m} e^{2\pi i s j/m} \sum_{k \in \mathcal{R}_n} e^{2\pi i s k/n}$$

$$= \sum_{\substack{j \in \mathcal{R}_m \\ k \in \mathcal{R}_n}} e^{2\pi i s (j/m + k/n)}$$

$$= \sum_{\substack{j \in \mathcal{R}_m \\ k \in \mathcal{R}_n}} e^{2\pi i s (jn + km)/mn},$$

but by Theorem 13.2, this is just $c_s(mn)$, so $c_s \in \mathcal{M}$.

Taking the sum of exponentials over somewhat larger ranges than the type we have been considering, we get a geometric series. For example,

$$\sum_{k=1}^{n} e^{2\pi i s k/n} = \begin{cases} e^{2\pi i s/n} \dfrac{\{e^{2\pi i s/n}\}^n - 1}{e^{2\pi i s/n} - 1} & \text{if } e^{2\pi i s/n} \neq 1, \\[2mm] n & \text{if } e^{2\pi i s/n} = 1. \end{cases}$$

But $e^{2\pi i s/n} = \cos(2\pi s/n) + i \sin(2\pi s/n) = 1$ if and only if $n|s$. From (8) and (9), $\{e^{2\pi i s/n}\}^n = e^{2\pi i s} = 1$. Thus, we have

$$\sum_{k=1}^{n} e^{2\pi i s k/n} = \begin{cases} 0, & n \nmid s, \\ n, & n|s. \end{cases}$$

Furthermore, if n is a prime power, $n = p^\beta$, then

$$\mathscr{R}_{p^\beta} = \{k : 1 \le k \le p^\beta\} - \{tp : 1 \le t \le p^{\beta-1}\}.$$

That is, the numbers in $\mathscr{R}_{p^\beta}$ are just the numbers from 1 to p^β whose gcd with p^β is not larger than 1; the numbers whose gcd with p^β is larger than 1 are all of the form tp, $1 \le t \le p^{\beta-1}$.

We write $p^\beta \| s$ (read "p^β is an *exact divisor* of s") to mean $p^\beta | s$ and $p^{\beta+1} \nmid s$. Then

$$c_s(p^\beta) = \sum_{k \in \mathscr{R}_{p^\beta}} e^{2\pi i s k / p^\beta}$$

$$= \sum_{k=1}^{p^\beta} e^{2\pi i s k / p^\beta} - \sum_{t=1}^{p^{\beta-1}} e^{2\pi i s t p / p^\beta}$$

$$= \begin{cases} 0, & p^\beta \nmid s \\ p^\beta, & p^\beta | s \end{cases} - \begin{cases} 0, & p^{\beta-1} \nmid s \\ p^{\beta-1}, & p^{\beta-1} | s \end{cases}$$

$$= \begin{cases} 0, & p^{\beta-1} \nmid s \\ -p^{\beta-1}, & p^{\beta-1} \| s \\ p^\beta - p^{\beta-1}, & p^\beta | s. \end{cases}$$

It is now easy to prove Theorem 14.3.

Theorem 14.3.

$$\sum_{d|n} c_s(d) = \begin{cases} n, & n|s \\ 0, & n \nmid s \end{cases}$$

Proof. It suffices to prove this theorem for $n = p^\beta$, prime p. If $p^\beta | s$, then $c_s(d) = \varphi(d)$ for every $d | p^\beta$ and from Theorem 10.1 it follows that $\Sigma c_s(d) = p^\beta$. If $p^\beta \nmid s$, say $p^\alpha \| s$, $0 \le \alpha < \beta$, and

$$\sum_{d|p^\beta} c_s(d) = c_s(1) + \sum_{j=1}^{\alpha} c_s(p^j) + \sum_{j=\alpha+1}^{\beta} c_s(p^j)$$

$$= 1 + \sum_{j=1}^{\alpha} (p^j - p^{j-1}) + (-p^{(\alpha+1)-1})$$

$$= 0.$$

Corollary 14.3.

$$c_s(n) = \sum_{\substack{d|n \\ d|s}} \mu\left(\frac{n}{d}\right) d$$

Proof. Let $\delta_s \in \mathscr{A}$ be defined by $\delta_s(n) = 0$ if $n \nmid s$, $= n$ if $n \mid s$. Theorem 14.3 says $c_s \cdot \iota_0(n) = \delta_s(n)$. By Möbius inversion, $c_s(n) = \mu \cdot \delta_s(n)$. Therefore,

$$c_s(n) = \sum_{d \mid n} \mu\left(\frac{n}{d}\right) \delta_s(d) = \sum_{\substack{d \mid n \\ d \mid s}} \mu\left(\frac{n}{d}\right) d.$$

EXERCISES

14-1. Find $c_1(n)$ for every $n \in \mathscr{Z}^+$. (*Hint:* Use Corollary 14.3.)

14-2. Prove $\delta_s \in \mathscr{M}$.

14-3. Prove $c_n \cdot \iota_1'(n) = \mu(n)$ for every $n \in \mathscr{Z}^+$.

14-4. Prove the set $\mathscr{T}$ (defined in the proof of Theorem 14.1) is a RRS (mod n).

14-5. Prove that

$$c_s(n) = \sum_{d \mid (n,s)} \mu\left(\frac{n}{d}\right) d.$$

14-6. Is c_s completely multiplicative? Is δ_s completely multiplicative?

14-7. Prove
  (a) $c_s \cdot \tau = \delta_s \cdot \iota_0$;
  (b) $c_s \cdot \sigma = \delta_s \cdot \iota_1$.

15. Wilson's Theorem

In this section we will prove the result known as Wilson's Theorem: The integer $n > 1$ is prime if and only if $(n-1)! \equiv -1$ (mod n). Though this is not a practical test for primality for large values of n, the result is of theoretical interest and arises as a natural consequence of a theorem of Lagrange, which is proved below.

Theorem 15.1. (*Lagrange.*) *Suppose p is a prime and*

$$f(x) = \sum_{k=0}^{n} c_k x^k$$

is a polynomial with integral coefficients. Then either the congruence $f(x) \equiv 0$ (mod p) has at most n incongruent solutions or $c_k \equiv 0$ (mod p) for $k = 0, 1, \ldots, n$.

Proof. We prove this theorem by induction on the degree of $f(x)$. If $n = 1$ and $c_1 \not\equiv 0$ (mod p), then there is just one solution of $f(x) \equiv 0$ (mod p), by Corollary 12.2. Suppose Lagrange's Theorem is true for polynomials of degree $\leq n - 1$; assume that $f(x)$ is of degree n, and the congruence $f(x) \equiv 0$ (mod p) has at least $n + 1$ solutions $a_1, \ldots, a_n, a_{n+1}$ which are incongruent (mod p).

If $c_n \equiv 0 \pmod{p}$, we construct the polynomial

$$g(x) = \sum_{k=0}^{n-1} c_k x^k;$$

clearly all $n + 1$ solutions of $f(x) \equiv 0 \pmod{p}$ are also solutions of $g(x) \equiv 0 \pmod{p}$, and since $g(x)$ is of degree $n - 1$, we have contradicted the inductive supposition. Therefore, $c_n \not\equiv 0 \pmod{p}$.

Dividing $f(x)$ by $x - a_{n+1}$, we find

$$f(x) = (x - a_{n+1}) \, q(x) + r$$

where $r \in \mathscr{Z}$ and $q(x)$ is a polynomial with integral coefficients, of degree $n - 1$, and with leading coefficient c_n. Since

$$0 \equiv f(a_{n+1}) = (a_{n+1} - a_{n+1}) \, q(a_{n+1}) + r \equiv r \pmod{p},$$

we have $r \equiv 0$ and $f(x) \equiv (x - a_{n+1})q(x) \pmod{p}$; that is, the polynomial $f(x)$ and the polynomial $(x - a_{n+1})q(x)$ have corresponding coefficients which are congruent $\pmod{p}$. For every $k \neq n + 1$, $0 \equiv f(a_k) \equiv (a_k - a_{n+1})q(a_k) \pmod{p}$; since p is prime, and $a_k - a_{n+1} \not\equiv 0$, we have by Theorem 4.1 that $q(a_k) \equiv 0 \pmod{p}$ for $k = 1, \ldots, n$. Therefore, $q(x) \equiv 0 \pmod{p}$ is a polynomial congruence of degree $n - 1$ and has at least n incongruent solutions, which is a contradiction. This concludes the proof.

Notice that the theorem of Lagrange is not true if the modulus is not prime. For example, the congruence $x^2 - 5x \equiv 0 \pmod{6}$ has four solutions, $x = 0, 2, 3, 5$. Also the number of solutions may be strictly less than the degree, as in $x^3 - 2 \equiv 0 \pmod{5}$, which has only one solution, $x = 3$.

Theorem 15.2. (*Wilson.*) *A necessary and sufficient condition that $n > 1$ be prime is that $(n - 1)! \equiv -1 \pmod{n}$.*

Proof. Suppose n is an *odd* prime (if $n = 2$, the theorem is obvious). Consider the two polynomials

$$f(x) = x^{n-1} - 1$$

$$g(x) = (x - 1)(x - 2)\cdots(x - (n - 1))$$

and define $h(x) = f(x) - g(x)$. Clearly $h(x)$ has degree $n - 2$. The congruence $g(x) \equiv 0 \pmod{n}$ is satisfied by each of the numbers $1, 2, \ldots, n - 1$, and by Euler's Theorem these numbers are also solutions of $f(x) \equiv 0 \pmod{n}$. Therefore $h(x) \equiv 0 \pmod{n}$ has at least $n - 1$ solutions, so the coefficients are all $0 \pmod{n}$. In particular, the constant term in $h(x)$ is $0 \pmod{n}$, or

$$-1 - (-1)^{n-1}(n - 1)! \equiv 0 \pmod{n},$$

and this proves the necessity of the condition.

Conversely, if n is composite it has a prime divisor p, and p is a factor of $(n - 1)!$, or $(n - 1)! \equiv 0 \pmod{p}$. Evidently, then, by Exercise 11-3, $(n - 1)! \not\equiv -1 \pmod{n}$.

EXERCISES

15-1. Prove that if n is composite, then $(n - 1)! \equiv 0 \pmod{n}$ except for $n = 4$.

15-2. Consider again the polynomial $h(x)$ defined in the proof of Wilson's Theorem. Let b_k denote the coefficient of x^k in $h(x)$, for $k = 1, 2, \ldots, n - 2$. We proved that $b_k \equiv 0 \pmod{n}$. Describe the numbers b_k.

15-3. If $f(x) = g(x) h(x)$, prove that every solution of $f(x) \equiv 0 \pmod{p}$, prime p, is either a solution of $g(x) \equiv 0 \pmod{p}$ or of $h(x) \equiv 0 \pmod{p}$. Show by an example that this need not be true if the modulus is composite.

15-4. Suppose $f(x) = g^n(x)$, $n \in \mathscr{Z}^+$. If p is a prime, prove that every solution of $f(x) \equiv 0 \pmod{p^n}$ is a solution of $g(x) \equiv 0 \pmod{p}$, and vice versa.

15-5. Prove that $[(n - 1)!/n]$ is even for $n > 4$.

15-6. Show that Fermat's Theorem and Wilson's Theorem together imply and are implied by the following statement: For all $a \in \mathscr{Z}$ and all primes p, $a^p + a(p - 1)!$ is a multiple of p.

15-7. Show that the expression $a^p + a(p - 1)!$ in Exercise 15-6 may be replaced by $a + a^p(p - 1)!$.

16. Primitive Roots

Definition. *If $(a, m) = 1$, then a is said to* belong to *(the exponent) t (mod m) if and only if t is the least positive integer x such that $a^x \equiv 1 \pmod{m}$.*

Clearly, if a belongs to $t \pmod{m}$, then $1 \le t \le \varphi(m)$. In fact, t is a divisor of $\varphi(m)$ (see Exercise 13-3). More generally, suppose a belongs to $t \pmod{m}$. Then $a^x \equiv 1 \pmod{m}$ if and only if $t | x$.

Theorem 16.1. *If a belongs to t_1 (mod m_1) and a belongs to t_2 (mod m_2), then a belongs to $\langle t_1, t_2 \rangle$ (mod $\langle m_1, m_2 \rangle$).*

Proof. Say a belongs to $t \pmod{\langle m_1, m_2 \rangle}$ and let $T = \langle t_1, t_2 \rangle$. Now $a^t \equiv 1 \pmod{\langle m_1, m_2 \rangle}$ implies that $a^t \equiv 1 \pmod{m_1}$ and $t_1 | t$. Similarly, $t_2 | t$, so $T | t$. But $a^T \equiv 1 \pmod{m_1}$ and $a^T \equiv 1 \pmod{m_2}$ imply that $a^T \equiv 1 \pmod{\langle m_1, m_2 \rangle}$ by Exercise 11-9, so $t | T$.

Theorem 16.2. *If a belongs to t (mod m), then a^u belongs to $t/(t, u)$ (mod m).*

Proof. Suppose a^u belongs to $s \pmod{m}$, and let $d = (t, u)$. Since a belongs to t, $a^{us} \equiv 1$ implies $t | us$ and $(t/d) | s$. Similarly, since a^u belongs to s, $a^{u(t/d)} \equiv 1$ implies $s | (t/d)$.

Definition. *If a belongs to $\varphi(m)$ (mod m), then a is called a* primitive root *of m.*

Example. It is easy to see that 2 is a primitive root of 5, because $2^2 \equiv 4$, $2^3 \equiv 3$, $2^4 \equiv 1$ (mod 5) and $\varphi(5) = 4$. Also, 3 belongs to 4 (mod 5) so 3 is also a primitive root of 5. However, 1 belongs to 1 (mod 5) and 4 belongs to 2 (mod 5), so neither 1 nor 4 is a primitive root of 5.

Example. Not every $m > 1$ has primitive roots. For instance, consider $m = 8$. Evidently, if a is even, then $a^x \equiv 1$ (mod 8) has no solution by Exercise 13-3, so no even integer can be a primitive root of 8. If a is odd, then either $a \equiv \pm 1$ or $a \equiv \pm 3$ (mod 8); in any case, $a^2 \equiv 1$ (mod 8), so every odd integer belongs to 2 (mod 8) and hence is not a primitive root, since $\varphi(8) = 4$.

In view of the last two examples it is natural to ask what values of m have primitive roots. We will now concern ourselves with proving that the complete answer to this question is: The integer $m > 1$ has a primitive root if and only if $m = 2, 4, p^\beta$, or $2p^\beta$, where p is an odd prime.

Theorem 16.3. *Suppose p is an odd prime and $t|(p - 1)$. Then there are exactly $\varphi(t)$ incongruent numbers which belong to t (mod p).*

Proof. Let $\psi(t)$ be the number of incongruent numbers which belong to t (mod p). We recall that if there is an a which belongs to t (mod p), then $t|\varphi(p)$; contrapositively, if $t \nmid (p - 1)$, then $\psi(t) = 0$. Since there are exactly $p - 1$ incongruent numbers which are prime to p, and each of these belongs to some unique exponent (mod p), we have

(11) $$p - 1 = \sum_{t|(p-1)} \psi(t).$$

For each $t|(p - 1)$, $\psi(t) \geq 0$. If $\psi(t) \neq 0$, there is some a which belongs to t (mod p). Consider the congruence $x^t \equiv 1$ (mod p). Evidently each of the incongruent integers $a, a^2, \ldots, a^t$ is a solution, and by Lagrange's Theorem there can be no others. Hence every number which belongs to t is of the form a^u, $1 \leq u \leq t$. But by Theorem 16.2, a^u belongs to $t/(u,t)$, so a^u belongs to t if and only if $(u,t) = 1$, $1 \leq u \leq t$. There are clearly $\varphi(t)$ such numbers. Hence, for every $t|(p - 1)$, either $\psi(t) = 0$ or $\psi(t) = \varphi(t)$. From (11) and Theorem 10.1, we get

$$p - 1 = \sum_{t|(p-1)} \psi(t) \leq \sum_{t|(p-1)} \varphi(t) = p - 1.$$

Therefore for every $t|(p - 1)$, $\psi(t) = \varphi(t)$, and the proof is complete.

Corollary 16.3. *If p is an odd prime, there are $\varphi(\varphi(p)) = \varphi(p - 1)$ primitive roots of p. In particular, since $\varphi(p - 1) \geq 1$, every odd prime has a primitive root.*

We now introduce a function $\lambda \in \mathscr{A}$ and show that for every a prime to m, $a^{\lambda(m)} \equiv 1 \pmod{m}$, that is, $\lambda(m)$ is a *universal exponent* for m. This will tell us that if $\lambda(m) < \varphi(m)$, then m cannot have primitive roots.

$$\lambda(m) = \begin{cases} \varphi(m) & \text{if } m = 1,2,4,p^\beta,2p^\beta \ (p \text{ odd prime}); \\ \tfrac{1}{2}\varphi(m) & \text{if } m = 2^\beta, \beta \geq 3; \\ \langle \lambda(2^\beta), \lambda(p_1^{\beta_1}), \ldots, \lambda(p_r^{\beta_r}) \rangle & \text{if } m = 2^\beta \prod_{i=1}^r p_i^{\beta_i}, p_i \text{ odd primes.} \end{cases}$$

Theorem 16.4. *For every* $(a,m) = 1$, $a^{\lambda(m)} \equiv 1 \pmod{m}$.

Proof. If $m = 2, 4, p^\beta$, or $2p^\beta$ the result is immediate from Euler's Theorem. Suppose $m = 2^\beta$, $\beta \geq 3$. If $(a,2^\beta) = 1$, then a is odd and we have already observed that $a^2 \equiv 1 \pmod 8$, or that

$$a^{\lambda(2^3)} \equiv 1 \pmod{2^3}.$$

Suppose we have shown that for $3 \leq \beta - 1$,

$$a^{\lambda(2^{\beta-1})} \equiv 1 \pmod{2^{\beta-1}}.$$

Then for some k,

$$a^{\lambda(2^{\beta-1})} = 1 + k2^{\beta-1}.$$

Squaring both sides, we have

$$a^{2\lambda(2^{\beta-1})} = 1 + k2^\beta + k^2 2^{2\beta-2}$$

so that $a^{\lambda(2^\beta)} \equiv 1 \pmod{2^\beta}$.

Finally, suppose that

$$m = 2^\beta \prod_{i=1}^r p_i^{\beta_i}.$$

If $(a,m) = 1$, then $1 = (a,2^\beta) = (a,p_1^{\beta_1}) = \cdots = (a,p_r^{\beta_r})$, and we have already shown that

$$a^{\lambda(2^\beta)} \equiv 1 \pmod{2^\beta}$$

$$a^{\lambda(p_1^{\beta_1})} \equiv 1 \pmod{p_1^{\beta_1}}$$

$$\ldots$$

$$a^{\lambda(p_r^{\beta_r})} \equiv 1 \pmod{p_r^{\beta_r}}.$$

But $\lambda(2^\beta) | \lambda(m)$, and for $i = 1, \ldots, r$, $\lambda(p_i^{\beta_i}) | \lambda(m)$, so

$$a^{\lambda(m)} \equiv 1 \pmod{2^\beta}$$

$$a^{\lambda(m)} \equiv 1 \pmod{p_1^{\beta_1}}$$

$$\ldots$$

$$a^{\lambda(m)} \equiv 1 \pmod{p_r^{\beta_r}}.$$

Hence, each of the above moduli is a divisor of $a^{\lambda(m)} - 1$, and since the moduli are relatively prime in pairs, their product m is a divisor of $a^{\lambda(m)} - 1$. This completes the proof.

Corollary 16.4. *If m is a multiple of any number of the type* (a) $4p$, *where p is an odd prime, or* (b) pq, *where $p \neq q$ are odd primes, or* (c) 8, *then m cannot have a primitive root.*

Proof. Suppose

$$m = 2^\beta \prod_{i=1}^{r} p_i^{\beta_i}.$$

If m has a divisor of type (a), ($\beta \geq 2$ and $r \geq 1$), or of type (b), ($r \geq 2$), then

$$\lambda(m) = \langle \lambda(2^\beta), \ldots, \lambda(p_r^{\beta_r}) \rangle$$

$$< \lambda(2^\beta) \cdots \lambda(p_r^{\beta_r})$$

$$\leq \varphi(2^\beta) \cdots \varphi(p_r^{\beta_r}) = \varphi(m).$$

If $8|m$ ($\beta \geq 3$), again $\lambda(m) < \varphi(m)$. In any case, no a prime to m can belong to $\varphi(m)$ (mod m).

Lemma 16.5. *If a is a primitive root of p^β, then either a belongs to $\varphi(p^\beta)$ (mod $p^{\beta+1}$) or a is a primitive root of $p^{\beta+1}$. Every primitive root of $p^{\beta+1}$ is a primitive root of p^β.*

Proof. Suppose a belongs to t (mod $p^{\beta+1}$). Then $t|\varphi(p^{\beta+1})$. From $a^t \equiv 1$ (mod $p^{\beta+1}$) we have the same congruence (mod p^β), so $\varphi(p^\beta)|t$, since a belongs to $\varphi(p^\beta)$ (mod p^β). Therefore, either $t = \varphi(p^\beta)$ or $t = p\varphi(p^\beta) = \varphi(p^{\beta+1})$.

If a is a primitive root of $p^{\beta+1}$, and if a belongs to t (mod p^β), then $t|\varphi(p^\beta)$. Since $a^t \equiv 1$ (mod p^β), we have $a^t = 1 + kp^\beta$. Raising both sides of this equation to the power p, we get $a^{pt} \equiv 1$ (mod $p^{\beta+1}$). Therefore, $\varphi(p^{\beta+1})|pt$, so $t = \varphi(p^\beta)$ and a is a primitive root of p^β.

Theorem 16.5. *For every $\beta \in \mathscr{Z}^+$, there exists a primitive root of p^β, where p is an odd prime. If a is a primitive root of p, then at least one of a or $a + p$ is a primitive root of p^2. If a is a primitive root of p^β ($\beta \geq 2$), then a is a primitive root of $p^{\beta+1}$. Consequently, there exists a number a (independent of β) such that a is a primitive root of p^β for all $\beta \in \mathscr{Z}^+$.*

Proof. We know by Corollary 16.3 that there is a primitive root a of p. By Lemma 16.5, either a is a primitive root of p^2 or a belongs to $\varphi(p)$ (mod p^2). In the latter case, consider $a + p$, which is also a primitive root of p.

$$(a + p)^{\varphi(p)} = a^{\varphi(p)} + (p - 1)a^{\varphi(p)-1}p + \cdots$$

$$\equiv 1 - pa^{\varphi(p)-1} \pmod{p^2}.$$

Now if $a + p$ belongs to $\varphi(p)$, then $-pa^{\varphi(p)-1} \equiv 0 \pmod{p^2}$, or $a^{\varphi(p)-1} \equiv 0$ $\pmod{p}$, which is impossible. Therefore, $a + p$ belongs to $\varphi(p^2) \pmod{p^2}$, by Lemma 16.5.

Assume we have proved the existence of a primitive root for every p^x, $x = 1, \ldots, \beta$ ($\beta \geq 2$). If a is a primitive root of p^β, then by the lemma a is also a primitive root of $p^{\beta-1}$, so $a^{\varphi(p^{\beta-1})} = 1 + kp^{\beta-1}$, and $p \nmid k$. Raising both sides to the pth power we get

$$a^{p\varphi(p^{\beta-1})} \equiv 1 + kp^\beta \pmod{p^{\beta+1}}.$$

Since $p \nmid k$,

$$a^{p\varphi(p^{\beta-1})} = a^{\varphi(p^\beta)} \not\equiv 1 \pmod{p^{\beta+1}},$$

so by the lemma a is a primitive root of $p^{\beta+1}$. The induction is complete.

Corollary 16.5. *For every $\beta \in \mathscr{Z}^+$, there exists a primitive root of $2p^\beta$.*

Proof. We know there is a primitive root a of p^β. If q is even, then $a + p^\beta$ is an odd primitive root of p^β, so we may assume a is odd. Hence a is also a primitive root of 2. By Theorem 16.1, a belongs to $\langle \varphi(2), \varphi(p^\beta) \rangle = \varphi(2p^\beta)$ $\pmod{2p^\beta}$.

In those cases where m has primitive roots, namely for $m = 2, 4, p^\beta$, or $2p^\beta$, we can state exactly how many there are.

Theorem 16.6. *If m has a primitive root, then m has exactly $\varphi(\varphi(m))$ incongruent primitive roots.*

Proof. Suppose a is a primitive root of m. Then we can easily see (Exercise 16-4) that

$$\mathscr{S} = \{a^k : k = 1, 2, \ldots, \varphi(m)\}$$

is a RRS $\pmod{m}$. Consider the congruence

(12) $x^{\varphi(m)} \equiv 1 \pmod{m}$.

Every $a^k \in \mathscr{S}$ is a solution of (12). We know by Exercise 13-3 that all the solutions of (12) are in $\mathscr{S}$. By Theorem 16.2, the number of a^k, $1 \leq k \leq \varphi(m)$, which belong to $\varphi(m)$ is just the number of k, $1 \leq k \leq \varphi(m)$, such that $(k, \varphi(m)) = 1$. There are $\varphi(\varphi(m))$ such k.

We saw in Theorem 16.5 that every primitive root of p^β, $\beta \geq 2$, is also a primitive root of $p^{\beta+1}$. This does not hold for $\beta = 1$; for example, 8 is a primitive root of 3, but not of 3^2. In fact, for every odd prime p, there exists a number g such that g is a primitive root of p, but not of p^2. This can be seen easily from the last theorem. Let $g_1, \ldots, g_r$, $r = \varphi(\varphi(p))$, be the incongruent $\pmod{p}$ primitive roots of p. Then all the numbers $g_i + kp$ ($i = 1, \ldots, r$; $k = 0, 1, \ldots, p - 1$) are primitive roots of p and are incongruent $\pmod{p^2}$. But there are rp such numbers, and $rp > \varphi(\varphi(p^2))$, so not all of them can be primitive roots of p^2. This proves the following.

Corollary 16.6. *For every odd prime* p *there are* $\varphi(p - 1) = p\ \varphi(\varphi(p)) - \varphi(\varphi(p^2)) = p\ \varphi(p - 1) - \varphi(p^2 - p)$ *numbers incongruent* $(mod\ p^2)$ *which are primitive roots of* p, *but not of* p^2.

EXERCISES

16-1. Find at least one primitive root for each of the following moduli: 7, 9, 15, 17, 25, 27.

16-2. Suppose $(a,m) = 1$ and consider the sequence $\{b_1, b_2, b_3, \ldots\}$ where b_i is the least positive residue $(mod\ m)$ of a^i. Prove this sequence is periodic with period equal to the exponent to which a belongs $(mod\ m)$.

16-3. Suppose a belongs to t_1 $(mod\ m_1)$ and a belongs to t_2 $(mod\ m_2)$. To what exponent $(mod\ \langle m_1, m_2 \rangle)$ does a^v belong?

16-4.* If a is a primitive root of m, prove $\{a, a^2, \ldots, a^{\varphi(m)}\}$ is a RRS $(mod\ m)$.

16-5. Prove that if a is a primitive root of $p^{\beta + 1}$ ($\beta \geq 2$), then so is $a + kp^\beta$ for every k.

16-6. If $\beta \geq 2$ and $\mathscr{G}$ is a set of all the incongruent primitive roots of p^β, prove that the set

$$\mathscr{H} = \{a + kp^\beta : a \in \mathscr{G}, k = 0, 1, \ldots, p - 1\}$$

is a set of all the incongruent primitive roots of $p^{\beta + 1}$.

16-7. Let p be an odd prime and let $(a, p) = 1$. Consider the sequence $\{t_1, t_2, t_3, \ldots\}$ where a belongs to t_x $(mod\ p^x)$. Describe the sequence $\{t_x\}$.

16-8. Find an integer which is a primitive root of both 29 and 73, using the information that 2 is a primitive root of 29 and 5 is a primitive root of 73.

17. Quadratic Residues

We will now study polynomial congruences of the type

$$(13) \qquad f(y) = Ay^2 + By + C \equiv 0\ (mod\ p), (A,p) = 1,$$

Since some special techniques are required to deal with the prime 2, *throughout this section* p *will always denote an odd prime*. We first observe that it suffices to be able to solve congruences of the type

$$(14) \qquad\qquad x^2 \equiv a\ (mod\ p)$$

in order to find all the solutions of (13).

In (14), take $a \equiv B^2 - 4AC\ (mod\ p)$. If x_1 is a solution of (14), then there corresponds a unique y_1 such that $2Ay_1 \equiv x_1 - B$, since $(2A, p) = 1$. Now we have

$$0 \equiv x_1^2 - a \equiv (2Ay_1 + B)^2 - (B^2 - 4AC) \equiv 4Af(y_1)\ (mod\ p).$$

But $(4A,p) = 1$, so we have found a solution $y = y_1$ of (13). Conversely, if there are no solutions of (14) for $a \equiv B^2 - 4AC$, then there can be no solution of (13).

Definition. *Suppose $(a,p) = 1$. We say a is a* quadratic residue *of p if the congruence* (14) *is solvable. If* (14) *is not solvable, we say a is a* quadratic non-residue *of p.*

If x is a solution of (14), then $p - x$ is also a solution and $x \not\equiv p - x \pmod{p}$, so by Lagrange's Theorem these are all the solutions. Consider the numbers $1^2, 2^2, \ldots, \{(p - 1)/2\}^2$ which are obviously all quadratic residues of p. These are all distinct (mod p), because if $1 \le a < b \le (p - 1)/2$ and $a^2 \equiv b^2$ (mod p), then $(a - b)(a + b) = a^2 - b^2 \equiv 0 \pmod{p}$ and either $p|(a - b)$ or $p|(a + b)$; but both are impossible. If $(p - 1)/2 < a \le p - 1$, then $a^2 \equiv (p - a)^2$ and $1 \le p - a \le (p - 1)/2$. Thus all the quadratic residues of p are the distinct numbers $1^2, 2^2, \ldots, \{(p - 1)/2\}^2$. This proves Theorem 17.1.

Theorem 17.1. *There are exactly $(p - 1)/2$ quadratic residues and $(p - 1)/2$ quadratic non-residues of p.*

We will show next that a is a residue or a non-residue according as $a^{(p - 1)/2}$ is congruent to 1 or -1 (mod p). The statement of the result is facilitated by defining the Legendre symbol.

Definition. *Suppose $(a,p) = 1$. The* Legendre symbol (a/p) *is defined to be $+1$ if a is a quadratic residue of p, and -1 if a is a non-residue of p.*

Theorem 17.2. (*Euler's criterion.*) *If $(a,p) = 1$, then $a^{(p - 1)/2} \equiv (a/p) \pmod{p}$. Hence, a is a residue of p if and only if $a^{(p - 1)/2} \equiv 1 \pmod{p}$.*
  Proof. If $(a,p) = 1$, then

$$0 \equiv a^{p - 1} - 1 \equiv (a^{(p - 1)/2} - 1)(a^{(p - 1)/2} + 1) \pmod{p},$$

so $a^{(p - 1)/2} \equiv \pm 1 \pmod{p}$.
  Suppose $x^2 \equiv a \pmod{p}$ has a solution $x = x_1$. Then

$$a^{(p - 1)/2} \equiv (x_1^2)^{(p - 1)/2} \equiv 1 \pmod{p}.$$

Contrapositively, if $a^{(p - 1)/2} \equiv -1$, then a is a non-residue.
  Conversely, suppose $a^{(p - 1)/2} \equiv 1 \pmod{p}$, and let g be a primitive root of p. We know (Exercise 16-4) that the powers of g form a RRS (mod p), so there is some j, $1 \le j \le p - 1$, such that $g^j \equiv a \pmod{p}$. Then

$$g^{j(p - 1)/2} \equiv a^{(p - 1)/2} \equiv 1 \pmod{p}$$

so $\varphi(p) = (p - 1)|\{j(p - 1)/2\}$. Therefore j is even, say $j = 2J$. But then $g^j \equiv (g^J)^2 \equiv a \pmod{p}$, so that $(a/p) = 1$.

Corollary 17.2. *Suppose $(a,p) = 1$ and $(b,p) = 1$. Then $(a^2/p) = 1$, $(ab/p) = (a/p)(b/p)$, and if $a \equiv b \pmod{p}$, then $(a/p) = (b/p)$.*

Note. The equation $(ab/p) = (a/p)(b/p)$ says that the product of two residues is a residue, the product of two non-residues is a residue, and the product of a residue and a non-residue is a non-residue.

The problem of deciding if a given integer is a residue or non-residue of p is made easier by the properties of the Legendre symbol given in Corollary 17.2. In fact, it suffices to know (q/p) for all primes q in order to know (a/p) for all a. If $a > 1$ and $(a,p) = 1$, suppose $a = \Pi q_i^{\beta_i}$. Then $(a/p) = \Pi(q_i/p)^{\beta_i}$. Obviously, this has the same value as the simplified form

$$(a/p) = \prod_{\beta_i \, \text{odd}} (q_i/p).$$

If $a < -1$, then

$$(a/p) = (-1)^{(p-1)/2} \prod_{\beta_i \, \text{odd}} (q_i/p).$$

Example. Find $(-540/7)$. Since $-540 = -2^2 3^3 5$,

$$(-540/7) = (-1)^3(2/7)^2(3/7)^3(5/7) = -(3/7)(5/7).$$

But 1, 2, 4 are the residues of 7, so $(-540/7) = -1$. More simply, since $-540 \equiv -1 \pmod 7$, $(-540/7) = (-1/7) = (-1)^3 = -1$.

To be able to find (q/p) for prime q, we will use the following result.

Theorem 17.3. *(Gauss Lemma.) Suppose $(a,p) = 1$. Let u denote the number of the integers $a, 2a, \ldots, \{(p-1)/2\}a$ whose absolutely least residues $\pmod p$ are negative. Then $(a/p) = (-1)^u$.*

Proof. Let $r_1, \ldots, r_v$ be those residues such that $0 < r_i < p/2$ and $-s_1, \ldots, -s_u$ be those residues such that $-p/2 < -s_j < 0$. Obviously no two r_i are equal and no two s_j are equal. Also, no r_i can equal any s_j, because if so,

$$ma \equiv r_i = s_j \equiv -na \pmod p$$

for some m,n between 0 and $p/2$; but then $a(m + n) \equiv 0 \pmod p$, which is impossible since $a \not\equiv 0$ and $0 < m + n < p$. Therefore, the numbers $r_1, \ldots, r_v, s_1, \ldots, s_u$ are just the integers $1, 2, \ldots, (p-1)/2$ in some order, and

$$a(2a) \cdots \left(\frac{p-1}{2}a\right) \equiv r_1 \cdots r_v(-s_1) \cdots (-s_u) \equiv (-1)^u 1 \cdot 2 \cdots \frac{p-1}{2} \pmod p$$

$$a^{(p-1)/2}\left(\frac{p-1}{2}\right)! \equiv (-1)^u \left(\frac{p-1}{2}\right)! \pmod p.$$

Thus, $a^{(p-1)/2} \equiv (-1)^u \equiv (a/p) \pmod{p}$. But since $(-1)^u$ and (a/p) are ± 1, we have $(-1)^u = (a/p)$.

The proof of our next lemma is geometric, and a few introductory remarks will explain the notions involved. A *lattice point* in a Cartesian plane is a point with integer coordinates. If $\mathcal{R}$ is a bounded region in the plane, we let $\Lambda_{\mathcal{R}}$ denote the number of lattice points in $\mathcal{R}$. If $\mathcal{R}$ and $\mathcal{S}$ are bounded plane regions, then

(15) $$\Lambda_{\mathcal{R} \cup \mathcal{S}} = \Lambda_{\mathcal{R}} + \Lambda_{\mathcal{S}} - \Lambda_{\mathcal{R} \cap \mathcal{S}};$$

that is, the number of lattice points in $\mathcal{R} \cup \mathcal{S}$ is the number in $\mathcal{R}$ plus the number in $\mathcal{S}$ minus the number which have been counted twice (because they are both in $\mathcal{R}$ and in $\mathcal{S}$).

Recall that $[x]$ is the largest integer $\leq x$.

Lemma 17.4. *Suppose p and q are distinct odd primes, and define*

$$S(m,n) = \sum_{j=1}^{[(n-1)/2]} [jm/n].$$

Then $S(p,q) + S(q,p) = \frac{1}{2}(p-1) \cdot \frac{1}{2}(q-1)$.

Proof. For definiteness, assume $p > q$. In the plane, consider the three bounded regions $\mathcal{A}, \mathcal{B}, \mathcal{C}$ (see Figure 1) formed by the lines

$$L_1 : y = \frac{1}{2}$$

$$L_2 : y = \frac{q-1}{2}$$

$$L_3 : y = \frac{q}{p}x$$

$$L_4 : x = \frac{p-1}{2}$$

$$L_5 : x = \frac{1}{2}.$$

As in the figure, let $\mathcal{A}$ denote the region bounded by L_i, $i = 2,3,5$; $\mathcal{B}$, the region bounded by L_i, $i = 1,2,3,4$; and $\mathcal{C}$, the region bounded by L_i, $i = 2,3,4$.

Suppose j is an integer, $1 \leq j \leq (q-1)/2$, and consider the number of lattice points on the line $y = j$ and in $\mathcal{A}$. Clearly, these will be the points with coordinates $(1,j), (2,j), \ldots, (\zeta,j)$, where ζ is the largest possible abscissa of any point on this line $y = j$ which is not to the right of L_3. Since the lines

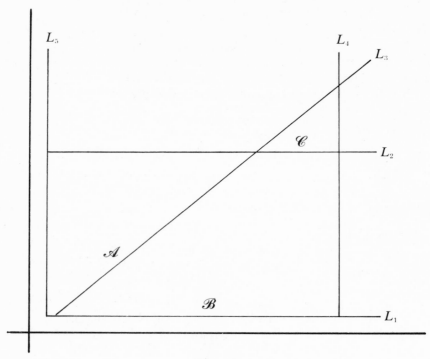

Figure 1

$y = j$ and L_3 intersect at $(jp/q, j)$, we see there are $\zeta = [jp/q]$ lattice points on $y = j$ and in $\mathscr{A}$. Consequently, the number $S(p,q)$ is just $\Lambda_{\mathscr{A}}$, and by similar arguments, $S(q,p) = \Lambda_{\mathscr{B} \cup \mathscr{C}}$, so

(16) $$S(p,q) + S(q,p) = \Lambda_{\mathscr{A}} + \Lambda_{\mathscr{B} \cup \mathscr{C}}.$$

Now we note that the only lattice points in $\mathscr{C}$ are on L_2. This may be seen by observing that

$$\left(\frac{p-1}{2}, \frac{q-1}{2} \right) = L_4 \cap L_2$$

is a lattice point and

$$\left(\frac{p-1}{2}, \frac{q(p-1)}{2p} \right) = L_3 \cap L_4$$

is less than 1 unit from $L_4 \cap L_2$. Therefore, every point in $\mathscr{C}$ is less than 1 unit from L_2. This shows that $\Lambda_{\mathscr{C}} = \Lambda_{\mathscr{B} \cap \mathscr{C}}$, and from (15) we get $\Lambda_{\mathscr{B} \cup \mathscr{C}} = \Lambda_{\mathscr{B}}$.

It is also easy to see that there are no lattice points on the segment of L_3 which is between L_1 and L_2, since p and q are distinct primes. Thus, $\Lambda_{\mathscr{A} \cap \mathscr{B}} = 0$, so that $\Lambda_{\mathscr{A} \cup \mathscr{B}} = \Lambda_{\mathscr{A}} + \Lambda_{\mathscr{B}}$. Making substitutions in (16) we have

$$S(p,q) + S(q,p) = \Lambda_{\mathscr{A} \cup \mathscr{B}}.$$

But obviously, $\Lambda_{\mathscr{A} \cup \mathscr{B}}$, the number of pairs (x,y) of integers satisfying $\frac{1}{2} \le x \le \frac{1}{2}(p-1)$ and $\frac{1}{2} \le y \le \frac{1}{2}(q-1)$, is $\frac{1}{2}(p-1) \cdot \frac{1}{2}(q-1)$. This completes the proof.

Theorem 17.4. (*Quadratic reciprocity law.*) *If p and q are distinct odd primes, then*

$$(p/q)(q/p) = (-1)^{\frac{1}{2}(p-1) \cdot \frac{1}{2}(q-1)}.$$

Proof. Suppose $1 \le k \le \frac{1}{2}(p-1)$. Then, from the division algorithm, $kq = p[kq/p] + t_k$, $0 < t_k \le p-1$. If $t_k < p/2$, then t_k is one of the r_i in the proof of Theorem 17.3, and if $p/2 < t_k$, then $t_k - p = -s_j$ for some j, $1 \le j \le u$. Recall that we showed in Theorem 17.3 that the r_i and the s_j are just the integers $1, 2, \ldots, (p-1)/2$ in some order. Hence

$$\frac{p^2-1}{8} = \sum_{k=1}^{(p-1)/2} k = \sum r_i + \sum s_j = \sum r_i + \sum_{t_k > p/2} (p - t_k)$$

$$= \sum r_i + up - \sum t_k,$$

or

(17) $$\sum r_i = \frac{p^2-1}{8} - up + \sum_{t_k > p/2} t_k.$$

Also,

$$\frac{q(p^2-1)}{8} = \sum_{k=1}^{(p-1)/2} kq = \sum_{k=1}^{(p-1)/2} \left\{ p\left[\frac{kq}{p}\right] + t_k \right\}$$

$$= pS(q,p) + \sum r_i + \sum_{t_k > p/2} t_k.$$

Combining this last equation with (17) we get

$$p(S(q,p) - u) = \frac{(q-1)(p^2-1)}{8} - 2\sum t_k.$$

But $(p^2-1)/8$ is an integer and $q-1$ is even, so

$$p(S(q,p) - u) \equiv 0 \,(\mathrm{mod}\ 2)$$

$$S(q,p) \equiv u \,(\mathrm{mod}\ 2).$$

Then $(-1)^{S(q,p)} = (-1)^u = (q/p)$. By a symmetric argument, $(-1)^{S(p,q)} = (p/q)$ and

$$(q/p)(p/q) = (-1)^{S(q,p) + S(p,q)} = (-1)^{\frac{1}{2}(p-1) \cdot \frac{1}{2}(q-1)}.$$

It is sometimes convenient to use the Law of Quadratic Reciprocity in the form

$$(p/q) = (-1)^{\frac{1}{2}(p-1) \cdot \frac{1}{2}(q-1)}(q/p).$$

To complete our knowledge we need to know the quadratic character of 2 modulo an odd prime. This is given by Theorem 17.5.

Theorem 17.5.

$$(2/p) = (-1)^{(p^2-1)/8} = \begin{cases} 1, & p \equiv \pm 1 \,(\text{mod } 8) \\ -1, & p \equiv \pm 3 \,(\text{mod } 8) \end{cases}$$

Proof. By the Gauss Lemma, $(2/p)$ will be $(-1)^u$, where u is the number of the integers $2, 2 \cdot 2, 3 \cdot 2, \ldots, \frac{1}{2}(p-1) \cdot 2$ which are larger than $p/2$. For $1 \le k \le (p-1)/2$, $2k > p/2$ if and only if $p/4 < k \le (p-1)/2$. The number of such k is $u = (p-1)/2 - [p/4]$.

We need only know if u is even or odd. If $p \equiv \pm 1 \,(\text{mod } 8)$, then $u \equiv 0 \,(\text{mod } 2)$, and if $p \equiv \pm 3 \,(\text{mod } 8)$, then $u \equiv 1 \,(\text{mod } 2)$. But also, if $p \equiv \pm 1 \,(\text{mod } 8)$, then

$$\frac{p^2 - 1}{8} \equiv 0 \,(\text{mod } 2),$$

and if $p \equiv \pm 3 \,(\text{mod } 8)$, then

$$\frac{p^2 - 1}{8} \equiv 1 \,(\text{mod } 2).$$

Example. Find $(60/23)$. We have

$$(60/23) = (14/23) = (2/23)(7/23) = (-1)^{(23^2-1)/8}(7/23) = (7/23)$$

$$= (-1)^{\frac{1}{2}(23-1) \cdot \frac{1}{2}(7-1)}(23/7) = -(2/7) = -(-1)^{(7^2-1)/8} = -1.$$

Or we could have computed $(60/23) = (2^2 \cdot 3 \cdot 5/23) = (3/23)(5/23)$, and then found $(3/23) = (-1)^{11 \cdot 1}(23/3) = -(2/3) = +1$, and $(5/23) = (-1)^{11 \cdot 2}(23/5) = (3/5) = -1$, so that again we find $(60/23) = -1$.

EXERCISES

17-1.   Find $(-540/11)$, $(540/11)$, $(311/19)$.

17-2.   Find two solutions of $x^2 \equiv 58 \,(\text{mod } 23)$.

17-3.   Prove Corollary 17.2.

17-4.* Suppose p is prime. Prove $(-3/p) = 1$ if $p \equiv 1 \,(\text{mod } 6)$, $(-3/p) = -1$ if $p \equiv 5 \,(\text{mod } 6)$.

17-5.   Prove: If p and q are distinct odd primes, then $(p/q) = (q/p)$ if and only if at least one of p,q is congruent to 1 (mod 4). Consequently, $(p/q) = -(q/p)$ if and only if $p \equiv q \equiv 3 \,(\text{mod } 4)$.

17-6.   Suppose $f(x) = ax^2 + bx + c$, a odd. Show that x_0 is a solution of

(18) $f(x) \equiv 0 \,(\text{mod } 2)$

if and only if x_0 is a solution of $(a + b)x + c \equiv 0 \,(\text{mod } 2)$. Hence, prove that if $b \equiv 0 \,(\text{mod } 2)$, (18) always has a solution, and if $b \equiv 1$ (mod 2), (18) has a solution if and only if $c \equiv 0 \,(\text{mod } 2)$. Also, if (18) has no roots, then $a \equiv b \equiv c \equiv 1 \,(\text{mod } 2)$.

17-7.   Prove that $(-1/p) \equiv p \,(\text{mod } 4)$ for all odd primes p.

18. Congruences with Composite Moduli

If $f(x)$ is a polynomial with integer coefficients and p is a prime, we will show that the solutions (if any) of $f(x) \equiv 0 \,(\text{mod } p^\beta)$ may be used to find all solutions of $f(x) \equiv 0 \,(\text{mod } p^{\beta+1})$.

Recall that Taylor's Theorem says if $f(x)$ is a polynomial of degree n, if $D^k f$ denotes the kth derivative of f (with the convention that $D^0 f = f$), then

$$f(x_0 + u) = \sum_{k=0}^{n} \frac{D^k f(x_0)}{k!} u^k.$$

Also, if

$$f(x) = \sum_{k=0}^{n} c_k x^k$$

and if the c_k are integers, then

$$\frac{D^i f(x_0)}{i!} = \sum_{k=i}^{n} \frac{k(k-1)\cdots(k-i-1)}{i!} c_k x_0^{k-i}$$

$$= \sum_{k=i}^{n} \binom{k}{i} c_k x_0^{k-i}$$

is an integer for each i, $0 \le i \le n$, whenever $x_0 \in \mathscr{Z}$. Therefore, if $x_0, u \in \mathscr{Z}$, then

$$f(x_0 + u) = \sum_{k=0}^{n} C_k u^k$$

where

$$C_k = \frac{D^k f(x_0)}{k!}$$

is an integer.

Theorem 18.1. *If $f(x)$ is a polynomial with integer coefficients, if p is a prime, and if x_0 is a solution of*

$$f(x) \equiv 0 \,(\text{mod } p^\beta),$$

then a necessary and sufficient condition that $x_0 + tp^\beta$ be a solution of

(19) $$f(x) \equiv 0 \,(\text{mod } p^{\beta + 1})$$

is that t be a solution of

(20) $$Df(x_0)t \equiv -\frac{f(x_0)}{p^\beta} \,(\text{mod } p).$$

Proof. Suppose

$$f(x) = \sum_{k=0}^{n} c_k x^k$$

and that $x_0 + tp^\beta$ is a solution of (19). By Taylor's Theorem,

$$f(x_0 + tp^\beta) = f(x_0) + Df(x_0)tp^\beta + \sum_{k=2}^{n} C_k(tp^\beta)^k \equiv 0 \,(\text{mod } p^{\beta + 1})$$

so that

$$f(x_0) + Df(x_0)tp^\beta \equiv 0 \,(\text{mod } p^{\beta + 1}).$$

But $p^\beta | f(x_0)$ by hypothesis, so

$$\frac{f(x_0)}{p^\beta} + Df(x_0)t \equiv 0 \,(\text{mod } p),$$

and the condition is necessary.

Conversely, suppose $f(x_0) \equiv 0 \,(\text{mod } p^\beta)$ and t is a solution of (20). But

$$f(x_0 + tp^\beta) \equiv f(x_0) + Df(x_0)tp^\beta \,(\text{mod } p^{\beta + 1}),$$

so $f(x_0 + tp^\beta) \equiv 0 \,(\text{mod } p^{\beta + 1})$ and the condition is sufficient.

Corollary 18.1. *Suppose x_0 is a solution of $f(x) \equiv 0 \,(\text{mod } p^\beta)$. Corresponding to x_0, there are $N = N(x_0)$ solutions $x_0 + tp^\beta$ of (19), where $N = 0$ if $p|Df(x_0)$ and $p\nmid\{f(x_0)/p^\beta\}$; $N = p$ if $p|Df(x_0)$ and $p|\{f(x_0)/p^\beta\}$; and $N = 1$ if $p\nmid Df(x_0)$.*
The proof is left as an exercise.

Example. Find all solutions of $x^3 + x + 3 \equiv 0 \,(\text{mod } 5^3)$. Clearly, $x_0 = 1$ is the only solution of $f(x) = x^3 + x + 3 \equiv 0 \,(\text{mod } 5)$; also, $Df(x_0) = (3x^2 + 1)|_{x=1} = 4$, $f(x_0)/5 = 1$. With N denoting the number of solutions of $4t \equiv -1 \,(\text{mod } 5)$, we have $N = 1$ $(t = 1)$. Now $x_0 + tp = 6$ is the only solution of $f(x) \equiv 0 \,(\text{mod } 5^2)$. With $x_1 = 6$, we find $Df(x_1) = 109$, $f(x_1)/5^2 = 9$, and N is the number of solutions of $109t \equiv -9 \pmod 5$, or $4t \equiv 1 \pmod 5$. Thus, $N = 1$ $(t = 4)$. Hence, $x_1 + tp^2 = 106$ is the unique solution of $f(x) \equiv 0 \pmod{5^3}$.

We now consider the general problem of solving a congruence

(21) $$f(x) \equiv 0 \pmod m$$

where

$$m = \prod_{i=1}^{r} p_i^{\beta_i}.$$

Evidently, if x_0 is a solution of (21), then x_0 is a solution of the system

(22) $$\begin{cases} f(x) \equiv 0 \pmod{p_1^{\beta_1}} \\ \quad \cdots \\ f(x) \equiv 0 \pmod{p_r^{\beta_r}}. \end{cases}$$

On the other hand, if x_i is a solution of $f(x) \equiv 0 \pmod{p_i^{\beta_i}}$, $i = 1, \ldots, r$, then by the Chinese Remainder Theorem there is a solution x_0 of the system

$$\begin{cases} x \equiv x_1 \pmod{p_1^{\beta_1}} \\ \quad \cdots \\ x \equiv x_r \pmod{p_r^{\beta_r}}; \end{cases}$$

then x_0, which is unique (mod m), is a solution of (22) and of (21). Therefore, all the solutions of (21) may be obtained by knowing the solutions of $f(x) \equiv 0$ (mod $p_i^{\beta_i}$) for each i, and we have seen that the solutions of these congruences may be constructed from the solutions of $f(x) \equiv 0 \pmod{p_i}$.

Example. Find all solutions of $f(x) = x^3 + 4x^2 + 3 \equiv 0 \pmod{78408}$. Notice that $78408 = 2^3 3^4 11^2$, and $Df(x) = 3x^2 + 8x$. Let $G_\beta(x,t) = f(x)/p^\beta + Df(x)t$.

We first find solutions of $f(x) \equiv 0 \pmod{2^3}$. Clearly, $a_0 = 1$ is the only solution of $f(x) \equiv 0 \pmod 2$, and $G_1(a_0,t) = 4 + 11t \equiv 0 \pmod 2$ has the single solution $t = 0$. Therefore, $a_0 + tp = a_1 = 1$ is the only solution of $f(x) \equiv 0 \pmod{2^2}$. $G_2(a_1,t) = 2 + 11t \equiv 0 \pmod 2$ has one solution, $t = 0$, so $a_1 + tp^2 = a_2 = 1$ is the only solution of $f(x) \equiv 0 \pmod{2^3}$.

Now, $f(x) \equiv 0 \pmod 3$ has solutions $b_0 = 0$ and $c_0 = 2$. $G_1(b_0,t) = 1 + 0 \cdot t \equiv 0 \pmod 3$ has no solution, so there are no solutions of $f(x) \equiv 0$ (mod 3^4) obtainable from b_0. But $G_1(c_0,t) = 9 + 28t \equiv 0 \pmod 3$ has just one solution, $t = 0$. Therefore, $c_0 + tp = c_1 = 2$ is the only solution of $f(x) \equiv 0 \pmod{3^2}$. $G_2(c_1,t) = 3 + 28t \equiv 0 \pmod 3$ has $t = 0$ for a solution, and $c_1 + tp^2 = c_2 = 2$ is the only solution of $f(x) \equiv 0 \pmod{3^3}$. $G_3(c_2,t) = 1 + 28t \equiv 0 \pmod 3$ has $t = 2$ for its solutions, and $c_2 + tp^3 = c_3 = 56$ is the only solution of $f(x) \equiv 0 \pmod{3^4}$.

Finally, $f(x) \equiv 0 \pmod{11}$ has solutions $d_0 = 3, e_0 = 6, g_0 = 9$. $G_1(d_0, t) = 6 + 51t \equiv 0 \pmod{11}$ has $t = 7$ for a solution, and this gives $d_1 = 80$ as a solution of $f(x) \equiv 0 \pmod{11^2}$. $G_1(e_0, t) = 33 + 156t \equiv 0 \pmod{11}$ has $t = 0$ as a solution, and $e_1 = 6$. $G_1(g_0, t) = 96 + 315t \equiv 0 \pmod{11}$ has $t = 2$ for its solution, so $g_1 = 31$.

We now need a solution of $x \equiv a_2 \pmod{2^3}$, $x \equiv c_3 \pmod{3^4}$, $x \equiv d_1 \pmod{11^2}$. By the Chinese Remainder Theorem, $x = 3^4 11^2 + 2^3 11^2 (20)(56) + 2^3 3^4 (76)(80) = 5{,}033{,}801 \equiv 15689 \pmod{78408}$ is the solution, and hence is a solution of $f(x) \equiv 0 \pmod{78408}$. Similarly, $y \equiv 1 \pmod{2^3}$, $y \equiv 56 \pmod{3^4}$, $y \equiv 6 \pmod{11^2}$ is solved by $y = 3^4 11^2 + 2^3 11^2 (20)(56) + 2^3 3^4 (76)(6) = 1{,}389{,}449 \equiv -21895$. And $z \equiv 1 \pmod{8}$, $z \equiv 56 \pmod{81}$, $z \equiv 31 \pmod{121}$ has the solution $z = 3^4 11^2 + 2^3 11^2 (20)(56) + 2^3 3^4 (76)(31) = 2{,}620{,}649 \equiv 33185$.

Therefore, the three solutions of $f(x) \equiv 0 \pmod{78408}$ are 15689, -21895, and 33185.

EXERCISES

18-1. Prove Corollary 18.1. [*Hint:* Use Corollary 12.2 to count the number of solutions of (20).]
18-2. Find all solutions of $x^3 + 2x - 3 \equiv 0 \pmod{125}$.
18-3. Find all solutions of $x^4 + x^3 + x + 1 \equiv 0 \pmod{2^2 3^3 7^2}$.
18-4. Show that if n is even, then $x^2 + x + 1 \equiv 0 \pmod{n}$ has no solution.
18-5. Show that if $5|n$, then $f(x) = x^6 + 3 \equiv 0 \pmod{n}$ has no solution. [*Hint:* Consider $f(x) \equiv 0 \pmod{5}$ and use Fermat's Theorem.]

Chapter 4

SUMMATORY FUNCTIONS

19. Introduction

The behavior of many of the arithmetic functions we have studied is quite erratic. For example, if p is prime, then $\tau(p) = 2$, and $\tau(p^k) = k + 1 \to \infty$ as $n = p^k \to \infty$. Since there are infinitely many primes, there are infinitely many $n \in \mathscr{Z}^+$ such that $\tau(n) = 2$; on the other hand, for every positive N there are infinitely many n such that $\tau(n) > N$. Thus, there is very little which can be said at this point about the behavior of $\tau(n)$ as n increases without bound. The situation is somewhat more encouraging if we consider the averaging function

$$T_1(n) = \frac{1}{n}\{\tau(1) + \cdots + \tau(n)\},$$

since averaging the values of the τ function will cause the function T_1 to change less drastically than does the τ function. But studying the function T_1 is equivalent to studying the function

$$T(n) = n \quad T_1(n) = \sum_{d=1}^{n} \tau(d).$$

Functions of the type $T(n)$ are called *summatory functions,* and in this chapter we study some of the summatory functions associated with the arithmetic functions we have seen previously. A few preliminaries are discussed first. Since $[x]$ is the largest integer not exceeding x, to every real x there corresponds a real number θ_x such that

$$x = [x] + \theta_x, \qquad 0 \le \theta_x < 1.$$

If $f \in \mathscr{A}$, we use the notation

$$\sum_{n \le x} f(n)$$

63

to mean

$$\sum_{n=1}^{[x]} f(n);$$

if $x < 1$, this null summation is defined to be zero.

Definition. *Let g be a function of a real or integral variable such that g(x) is defined and is positive for all x sufficiently large. The sets $\mathfrak{O}(g)$ and $\mathfrak{o}(g)$ are defined by*

$$\mathfrak{O}(g) = \left\{ f: \text{there exists a constant } M \text{ such that} \right.$$

$$\left. \frac{|f(x)|}{g(x)} < M \text{ for x sufficiently large} \right\},$$

$$\mathfrak{o}(g) = \left\{ f: \lim_{x \to \infty} \frac{f(x)}{g(x)} = 0 \right\}.$$

The elements of $\mathfrak{O}(g)$ are denoted $O(g)$ and the elements of $\mathfrak{o}(g)$ are denoted $o(g)$. We refer to both the class $\mathfrak{O}(g)$ and the elements $O(g)$ as "big-O g", and $\mathfrak{o}(g)$, $o(g)$ are both read "little-o g".
We write $f(x) \sim g(x)$ if and only if

$$\lim_{x \to \infty} \frac{f(x)}{g(x)} = 1;$$

the relation "$f(x) \sim g(x)$" is read "f is asymptotically equivalent to g."

For example, $\sin x = O(1)$; that is, $\sin x$ is an element of the class $\mathfrak{O}(1)$, since we can find a constant M, say $M = 2$, such that

$$\frac{|\sin x|}{1} < 2 \quad \text{for all } x.$$

As further examples, we have $x^2 = o(e^x)$, $\sin x = O(x)$, $\sin x = o(x)$, and $x^2 + 2x \sim x^2$.

Since both big-O and little-o terms will arise in equations we will be dealing with, we must assign some convention for combining such terms. Therefore we agree to write

$$O(f) = O(g)$$

if and only if $\mathfrak{O}(f) \subset \mathfrak{O}(g)$. *Notice that this is a non-symmetric use of the equality sign.* We have, for example, $O(x) = O(x^2)$, but *not* $O(x^2) = O(x)$. We also agree to denote by $O(f) \pm O(g)$ any element of the class $\mathfrak{O}(h)$ where

$$\mathfrak{O}(h) = \{u + v : u = O(f), v = O(g)\};$$

it should be clear that here $u + v$ is the *sum of functions* defined at each x in the common domain of u and v by $(u + v)(x) = u(x) + v(x)$. We define $O(f)\,O(g)$ to be any element of the class

$$\mathfrak{O}(h) = \{uv : u = O(f), v = O(g)\},$$

and uv is defined by $uv(x) = u(x)\,v(x)$. Similar rules hold for combining little-o terms.

As an immediate consequence of these definitions, we have $O(f) = O(g)$ if and only if $h = O(f)$ implies $h = O(g)$. Also, $O(f) = O(g)$ if and only if $f = O(g)$.

The following properties hold.

(1) $$O(g) \pm O(g) = O(g)$$

(2) $$o(g) \pm o(g) = o(g)$$

(3) $$O(O(g)) = O(g)$$

(4) $$O(o(g)) = o(g)$$

(5) $$\{O(g)\}^2 = O(g^2).$$

To prove (1), suppose $f = O(g)$ and $h = O(g)$. Then there are constants M_1 and M_2 such that $|f(x)| < M_1 g(x)$ and $|h(x)| < M_2 g(x)$, from which we have

$$|f(x) \pm h(x)| \le |f(x)| + |h(x)| < (M_1 + M_2)g(x),$$

so that $f \pm h = O(g)$.

Property (4) may be proved as follows. Let $f = o(g)$ and $h = O(f)$, so that $h = O(o(g))$. There is a constant M such that $|h(x)| < Mf(x)$ and $f(x)/g(x) \to 0$. Then

$$0 \le \lim_{x \to \infty} \frac{|h(x)|}{g(x)} \le M \lim_{x \to \infty} \frac{f(x)}{g(x)} = 0,$$

and $h = o(g)$.

It need not be true in general that

$$\sum_{i=1}^{n} O(g_i(x)) = o\left(\sum_{i=1}^{n} g_i(x)\right).$$

For example, for the sequence of constant functions $f_i(u) = i$, we have $f_1(u) = O(1)$, $f_2(u) = O(1)$, $f_3(u) = O(1), \ldots$, but

$$\sum_{i \le x} f_i(u) = \frac{[x]([x] + 1)}{2} = O(x^2) \ne o\left(\sum_{i \le x} 1\right) = O(x).$$

A sufficient condition is that the constants implied by the O symbols be uniformly bounded. With such a restriction we may add infinitely many error terms $O(g_i(x))$.

Suppose $|f_i(x)| < M_i g_i(x)$ for $i = 1, 2, \ldots$, that there exists a cónstant M such that $M_i < M$ for all i, and that

$$\sum_{i=1}^{\infty} g_i(x)$$

converges for every $x \geq x_0$. Then

(6)
$$\sum_{i=1}^{\infty} O(g_i(x)) = O\left(\sum_{i=1}^{\infty} g_i(x)\right).$$

This is because $|\Sigma f_i(x)| \leq \Sigma |f_i(x)| \leq M\Sigma g_i(x)$. Obviously, (6) also holds if the series are replaced by finite sums.

The above list of properties is certainly not exhaustive, but contains typical rules for combining terms involving the symbols O and o. It is usually much easier to derive whatever rules are necessary in some particular application than to memorize a lengthy list of such rules.

Lemma 19.1. *There is a constant γ, called Euler's constant, such that $1/2 < \gamma < 1$ and*

$$\sum_{n \leq x} \frac{1}{n} = \log x + \gamma + O\left(\frac{1}{x}\right).$$

Proof. We first prove the lemma for integers x, and then extend the result to the general case. Thus, assume $x = N$ is an integer. We define

$$\gamma_N = \sum_{n=1}^{N} \frac{1}{n} - \log N, \quad N = 1, 2, 3, \ldots.$$

We note that the sequence $\{\gamma_N\}$ is convergent, because

$$\gamma_N - \gamma_{N+1} = \log(N+1) - \log(N) - \frac{1}{N+1}$$

$$= \int_N^{N+1} \frac{dx}{x} - \frac{1}{N+1}.$$

Geometrically, this is the area of the plane region bounded by the curve $y = 1/x$, the x-axis, and the lines $x = N$ and $x = N + 1$ *minus* the area of a rectangle of length 1 and height $1/(N+1)$. Since a rectangle of this type may be inscribed in the region with area $\int_N^{N+1} x^{-1}\, dx$, the difference is positive (the shaded area in Figure 2). Hence, $\gamma_N - \gamma_{N+1} > 0$, so the sequence is decreasing.

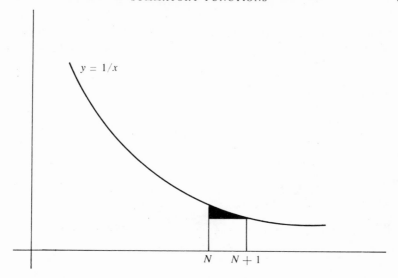

Figure 2

Furthermore, for every N, $\gamma_N > 0$, which can be seen as follows. We have

$$\gamma_N = 1 - \left\{ \log N - \sum_{n=2}^{N} \frac{1}{n} \right\}$$

$$= 1 - \left\{ \int_1^N \frac{dx}{x} - \sum_{n=2}^{N} \frac{1}{n} \right\}$$

$$= 1 - \left\{ \sum_{n=2}^{N} \int_{n-1}^{n} \frac{dx}{x} - \sum_{n=2}^{N} \frac{1}{n} \right\}$$

$$= 1 - \sum_{n=2}^{N} \left\{ \int_{n-1}^{n} \frac{dx}{x} - \frac{1}{n} \right\}.$$

Geometrically, this is the area of a unit square *minus* the sum of shaded areas of the type in Figure 2 for $n = 2, \ldots, N$. These shaded areas may be translated to the left by $n - 1$ units; since $y = 1/x$ is continuously decreasing, these areas $\int_{n-1}^{n} x^{-1} dx - 1/n$ will not overlap when translated into the unit square $0 \leq y \leq 1, 0 \leq x \leq 1$ and then γ_N is the (positive) area of the unshaded portion of the unit square. The argument is illustrated for γ_4 in Figure 3.

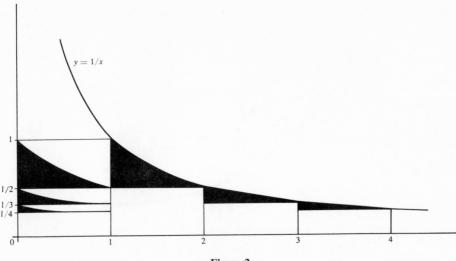

Figure 3

Since $\gamma_N > 0$ for all N and the sequence $\{\gamma_N\}$ is decreasing, it is convergent, say

$$\lim_{n \to \infty} \gamma_n = \gamma.$$

From our arguments above it is clear that γ is the limit of the unshaded region in the unit square under the translations described. Since this unshaded region is in the square, we evidently have $\gamma < 1$. Also, it is geometrically obvious that γ is the sum of a series whose N^{th} term a_N is the area of the unshaded portion of the rectangle $0 \le x \le 1$, $1/N \le y \le 1/(N + 1)$. Since this unshaded portion of the rectangle exceeds half the area of the rectangle (the shaded part of each rectangle lies entirely below the diagonal), we have

$$a_N > \frac{1}{2}\left(\frac{1}{N} - \frac{1}{N + 1}\right) = \frac{1}{2N(N + 1)}.$$

Therefore,

$$\gamma = \sum_{N=1}^{\infty} a_N > \frac{1}{2} \sum_{N=1}^{\infty} \frac{1}{N(N + 1)} = \frac{1}{2}.$$

Finally, we consider $\gamma_N - \gamma$. By the translation of shaded areas we know this is the area of the unshaded portion of the unit square inside the region $0 \le x \le 1$, $1/N \le y \le 1$, minus the limit of the unshaded areas. But this difference lies within the region $0 \le x \le 1$, $0 \le y \le 1/N$ with area $1/N$.

Hence $\gamma_N - \gamma < 1/N$, or $\gamma_N - \gamma = O(1/N)$; applying definitions, we have

$$\sum_{n=1}^{N} \frac{1}{n} = \log N + \gamma + O\left(\frac{1}{N}\right),$$

which is the lemma in case x is an integer.

Now suppose $x \geq 1$ is arbitrary (not necessarily an integer). By what we have already proved,

(7)
$$\sum_{n \leq x} \frac{1}{n} = \sum_{n=1}^{[x]} \frac{1}{n} = \log [x] + \gamma + O\left(\frac{1}{[x]}\right).$$

Since $[x] = x + O(1)$, we have

$$\log [x] = \log (x + O(1)) = \log \left(x\left\{1 + O\left(\frac{1}{x}\right)\right\}\right)$$

(8)
$$\log [x] = \log x + \log \left(1 + O\left(\frac{1}{x}\right)\right).$$

Now recall that for $|t| < 1$,

$$\log (1 + t) = \sum_{n=1}^{\infty} \frac{(-1)^{n-1} t^n}{n}$$

so that

$$\log \left(1 + O\left(\frac{1}{x}\right)\right) = \sum_{n=1}^{\infty} \frac{(-1)^{n-1}(O(1/x))^n}{n}$$

$$= \sum_{n=1}^{\infty} O(1/x^n)$$

$$= O\left(\sum_{n=1}^{\infty} \frac{1}{x^n}\right)$$

since the constant involved in $O(1/x^n)$ is θ_x^n/n, and these are uniformly less than 1. Therefore,

$$\log \left(1 + O\left(\frac{1}{x}\right)\right) = O\left(\frac{1}{x} \frac{x}{x-1}\right) = O\left(\frac{1}{x}\right).$$

Thus, from (8) we have $\log [x] = \log x + O(1/x)$. These results together with Exercise 19-2 into (7) give us

$$\sum_{n \leq x} \frac{1}{n} = \log x + O(1/x) + \gamma + O(1/x)$$

$$= \log x + \gamma + O(1/x).$$

The proof is complete.

Note. Lemma 19.1 also implies (in a trivial way) certain weaker asymptotic formulas. For example,

$$\sum_{n \le x} \frac{1}{n} = \log x + O(1),$$

because $\gamma = O(1)$ and $O(1/x) = O(1)$. In many applications, we lose nothing by using a weaker result of this type. For example, see the Note after Theorem 21.1.

EXERCISES

19-1. Prove properties (2), (3), and (5).

19-2. Show that $O(1/[x]) = O(1/x)$.

19-3. Prove that "$\sim$" is an equivalence relation on the set of positive-valued functions.

19-4. Prove the big-O relation is reflexive and transitive, but not symmetric, on the set of positive-valued functions.

19-5. Prove that the little-o relation is transitive. Show by examples that it is neither reflexive nor symmetric.

19-6.* Prove

$$\sum_{n \le x} \frac{1}{n} = O(\log x).$$

19-7. Prove the remarks made earlier, namely that the following conditions are equivalent.
  (a) $O(f) = O(g)$;
  (b) $h = O(f)$ implies $h = O(g)$;
  (c) $f = O(g)$.

19-8. Prove the following conditions are equivalent:
  (a) $o(f) = o(g)$;
  (b) if $h = o(f)$, then $h = o(g)$;
  (c) $f = o(g)$.

19-9. Prove that $O(O(O(g))) = O(g)$.

19-10. Prove $o(o(g)) = o(g)$.

20. The Euler–McLaurin Sum Formula

It is interesting to note that techniques similar to those employed in the proof of Lemma 19.1 may be applied in a more general setting to obtain a convenient tool for dealing with sums of the type under consideration in this chapter. While we will not need this result until later, we now prove Theorem 20.1.

Theorem 20.1. (*Euler–McLaurin sum formula.*) *Suppose* $f(x)$ *is continuously differentiable for* $x \geq 1$. *Then*

$$\sum_{n \leq x} f(n) = f(1) + \int_1^x f(t)\,dt + \int_1^x (t - [t])Df(t)\,dt - (x - [x])f(x).$$

Note. In many applications it is sufficient to replace the last term by $O(f)$.

Proof. We first prove the theorem when $x = N$ is an integer. Then

$$\sum_{n \leq N} f(n) = N\,f(N) - \sum_{n=2}^{N} (n-1)\{f(n) - f(n-1)\}$$

$$= N\,f(N) - \sum_{n=2}^{N} (n-1) \int_{n-1}^{n} Df(t)\,dt$$

$$= N\,f(N) - \sum_{n=2}^{N} \int_{n-1}^{n} [t]Df(t)\,dt$$

$$(9) \qquad \sum_{n \leq N} f(n) = N\,f(N) - \int_1^N [t]Df(t)\,dt.$$

But, using integration by parts,

$$\int_1^N f(t)\,dt = t\,f(t)\Big]_1^N - \int_1^N t\,Df(t)\,dt = N\,f(N) - f(1) - \int_1^N t\,Df(t)\,dt.$$

Substituting into (9) the expression for $N\,f(N)$ from this equation, we have

$$\sum_{n \leq N} f(n) = f(1) + \int_1^N f(t)\,dt + \int_1^N (t - [t])Df(t)\,dt,$$

which is the result in Theorem 20.1 when $x = [x]$ is an integer.

Now suppose x is arbitrary. Then

$$\sum_{n \leq x} f(n) = f(1) + \int_1^{[x]} f(t)\,dt + \int_1^{[x]} (t - [t])Df(t)\,dt$$

$$= f(1) + \int_1^x f(t)\,dt + \int_1^x (t - [t])Df(t)\,dt$$

$$- \left\{ \int_{[x]}^x f(t)\,dt + \int_{[x]}^x (t - [t])Df(t)\,dt \right\}.$$

But

$$\int_{[x]}^x f(t)\,dt + \int_{[x]}^x t\,Df(t)\,dt - \int_{[x]}^x [t]Df(t)\,dt$$

$$= t f(t)\Big]_{[x]}^x - [x] f(t)\Big]_{[x]}^x = (x - [x])f(x);$$

here we have used integration by parts and observed that for $[x] \leq t \leq x$, $[t] = [x]$ is constant. The proof is complete.

Example. Use Theorem 20.1 to prove there is a constant C_1 such that

$$\sum_{n \leq x} \frac{1}{n} = \log x + C_1 + O(1/x).$$

In Theorem 20.1, take $f(t) = 1/t$. Then $Df(t) = -1/t^2$ and

$$\sum_{n \leq x} \frac{1}{n} = 1 + \int_1^x \frac{dt}{t} - \int_1^x \frac{(t - [t])}{t^2} dt - \frac{x - [x]}{x}.$$

Clearly

$$\frac{x - [x]}{x} = O(1/x).$$

Also, if the following improper integrals are convergent, we may write

$$(10) \qquad \sum_{n \leq x} \frac{1}{n} = 1 + \log x - \int_1^\infty \frac{t - [t]}{t^2} dt + \int_x^\infty \frac{t - [t]}{t^2} dt + O\left(\frac{1}{x}\right).$$

But the first integral is convergent if the second approaches 0 as $x \to \infty$, and we see that this is the case since

$$0 \leq \int_x^\infty \frac{t - [t]}{t^2} dt \leq \int_x^\infty \frac{dt}{t^2} = \frac{1}{x}.$$

Therefore

$$\int_x^\infty \frac{t - [t]}{t^2} dt = O\left(\frac{1}{x}\right)$$

and

$$\int_1^\infty \frac{t - [t]}{t^2} dt$$

converges, say with value $1 - C_1$. Substitution into (10) proves the result.

EXERCISES

20-1.   We have defined

$$C_1 = 1 - \int_1^\infty \frac{t - [t]}{t^2} dt$$

$$\gamma = \lim_{n \to \infty} \left\{ \sum_{n=1}^N \frac{1}{n} - \log N \right\}.$$

Prove that $C_1 = \gamma$.

20-2. Suppose N is an integer, $1 \le N \le x$, and $f(t)$ is continuously differentiable for $t \ge 1$. Prove that

$$\sum_{n=N}^{x} f(n) = f(N) + \int_{N}^{x} f(t)\, dt + \int_{N}^{x} (t - [t]) Df(t)\, dt + O(f).$$

20-3.* Prove that

$$\sum_{n \le x} \frac{1}{n^2} = K - \frac{1}{x} + O(1/x^2)$$

for some constant K, $1 < K < 2$. It is known that $K = \pi^2/6$.

20-4. If $k \in \mathscr{Z}^{+}$, $k > 1$, prove

$$\sum_{n \le x} \frac{1}{n^k} = \frac{k}{k - 1} + O(x^{-k+1}).$$

20-5. Notice that $0 \le t - [t] < 1$ for all real t. Use this to show that

$$\int_{1}^{x} (t - [t])\, t^{k-1}\, dt = O(x^k)$$

for all $k \in \mathscr{Z}^{+}$. Hence use the Euler–McLaurin sum formula to prove that for every $k \in \mathscr{Z}^{+}$

$$\sum_{n \le x} n^k = C(k)\, x^{k+1} + O(x^k)$$

where $C(k)$ is a constant depending only on k, not on x.

21. Order of Magnitude of $\tau(n)$

Theorem 21.1.

$$\sum_{n \le x} \tau(n) = x \log x + O(x)$$

Proof. Let the plane region $\mathscr{A}$ be the region bounded by the lines $u = 1/2$, $v = 1/2$, and the hyperbola $uv = x$ (see Figure 4), where x is arbitrary but fixed. We will show that

$$\sum_{n \le x} \tau(n) = \Lambda_{\mathscr{A}} = \sum_{n \le x} [x/n].$$

If $1 \le n \le x$, then $\tau(n)$ is just the number of lattice points on the hyperbola $uv = n$, because corresponding to each positive divisor d of n, there is an integer n/d such that $(u,v) = (d, n/d)$ is a lattice point on $uv = n$. Thus, the sum $\sum_{n \le x} \tau(n)$ represents the number of lattice points in the first quadrant on any hyperbola $uv = n$, $1 \le n \le x$. But every lattice point in $\mathscr{A}$ is on exactly one such hyperbola, so $\sum_{n \le x} \tau(n) = \Lambda_{\mathscr{A}}$.

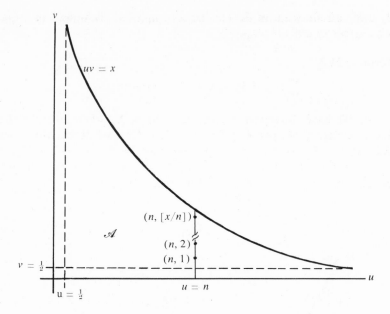

Figure 4

Again with $1 \leq n \leq x$, consider the number of lattice points in $\mathscr{A}$ and on the line $u = n$ (see Figure 4). These are just the points $(n,1), \ldots, (n,\xi)$, where ξ is the largest integer such that $n\xi \leq x$, or $\xi = [x/n]$. Since all lattice points in $\mathscr{A}$ are on exactly one such line, $\Lambda_{\mathscr{A}} = \Sigma_{n \leq x}[x/n]$. Therefore,

$$\sum_{n \leq x} \tau(n) = \sum_{n \leq x} [x/n]$$

$$= \sum_{n \leq x} \left\{ \frac{x}{n} + O(1) \right\}$$

$$= x \sum_{n \leq x} \frac{1}{n} + O\left(\sum_{n \leq x} 1 \right) = x \left\{ \log x + \gamma + O\left(\frac{1}{x} \right) \right\} + O(x)$$

$$= x \log x + \gamma x + O(1) + O(x) = x \log x + O(x).$$

Note. In the above, we could have used the weaker result given in the Note at the end of Section 19 and we would have obtained the same result. However, the still weaker form given in Exercise 19-6 could *not* have been used to obtain this same theorem.

By using a refinement of the techniques employed above we can prove a much sharper result.

Theorem 21.2.

$$\sum_{n \le x} \tau(n) = x(\log x + 2\gamma - 1) + O(\sqrt{x})$$

Proof. We have already shown that $\Sigma_{n \le x}\tau(n) = \Lambda_{\mathscr{A}}$ where $\mathscr{A}$ is the plane region described in the proof of Theorem 21.1. Now we divide $\mathscr{A}$ into sub-regions $\mathscr{B}$, $\mathscr{C}$, and $\mathscr{D}$ as follows: $\mathscr{B}$ is bounded by $v = 1/2, u = 1/2, u = [\sqrt{x}]$, $uv = x$; $\mathscr{C}$ is bounded by $u = 1/2, v = 1/2, v = [\sqrt{x}], uv = x$; $\mathscr{D}$ is bounded by $u = [\sqrt{x}], v = [\sqrt{x}], uv = x$. See Figure 5.

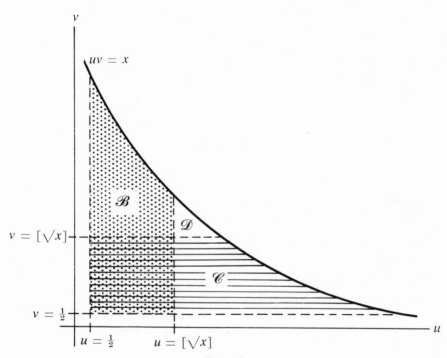

Figure 5

Evidently, $\Lambda_{\mathscr{A}} = \Lambda_{\mathscr{B}} + \Lambda_{\mathscr{C}} - \Lambda_{\mathscr{B} \cap \mathscr{C}} + \Lambda_{\mathscr{D}} = 2\Lambda_{\mathscr{B}} - \Lambda_{\mathscr{B} \cap \mathscr{C}} + \Lambda_{\mathscr{D}}$. Since $\mathscr{B} \cap \mathscr{C}$ is a square it is clear that $\Lambda_{\mathscr{B} \cap \mathscr{C}} = [\sqrt{x}]^2 = (\sqrt{x} + O(1))^2 = x + O(\sqrt{x})$. Also, $\mathscr{D} \cap \mathscr{C}$ is the line segment from $([\sqrt{x}],[\sqrt{x}])$ to $(x/[\sqrt{x}],[\sqrt{x}])$, with length $x/[\sqrt{x}] - [\sqrt{x}] < 3$; thus $\mathscr{D}$ may be inscribed in a square with sides of length 3, and $\Lambda_{\mathscr{D}}$ cannot exceed the area of such a square, so $\Lambda_{\mathscr{D}} = O(1)$.

Finally,

$$2\Lambda_{\mathscr{B}} = 2 \sum_{n \le \sqrt{x}} [x/n] = 2 \sum_{n \le \sqrt{x}} \left\{ \frac{x}{n} + O(1) \right\}$$

$$= 2x \sum_{n \le \sqrt{x}} \frac{1}{n} + O\left(\sum_{n \le \sqrt{x}} 1 \right)$$

$$= 2x(\log \sqrt{x} + \gamma + O(1/\sqrt{x})) + O(\sqrt{x})$$

$$= x \log x + 2\gamma x + O(\sqrt{x}).$$

Combining results we have

$$\sum_{n \le x} \tau(n) = x \log x + 2\gamma x + O(\sqrt{x}) - x - O(\sqrt{x}) + O(1)$$

$$= x(\log x + 2\gamma - 1) + O(\sqrt{x}).$$

The results in either of the last two theorems may be stated loosely by saying that $\tau(n)$ has average order of magnitude equal to $\log n$; that is,

$$\frac{1}{N} \sum_{n \le N} \tau(n) \sim \log N.$$

This relation follows easily from either theorem: since

$$T(N) = \sum_{n \le N} \tau(n) = N \log N + O(N),$$

we have $(1/N)T(N) = \log N + O(1)$, and

$$\lim_{N \to \infty} \frac{(1/N)T(N)}{\log N} = \lim_{N \to \infty} \frac{\log N + O(1)}{\log N} = 1.$$

22. Order of Magnitude of $\sigma(n)$

In the proofs of the last two theorems the same geometric argument was used to evaluate a function at the lattice points in a plane region in two different ways. The technique has a general applicability and may be used to prove the following.

Theorem 22.1.

$$\sum_{n \le x} f \cdot g(n) = \sum_{d \le x} f(d) \sum_{n \le x/d} g(n)$$

Proof.

$$\sum_{n \le x} f \cdot g(n) = \sum_{n \le x} \sum_{ab=n} f(a) g(b)$$

is the sum of the evaluations of $f(a)g(b)$ at each lattice point on the upper branch of $uv = n$ for every n, $1 \le n \le x$, so the evaluation extends over every lattice point in the region $\mathscr{A}$ of Figure 4.

Now let $d \in \mathscr{Z}$ be fixed, $1 \le d \le x$, and consider the sum of the evaluations of $f(d)g(n)$ at every lattice point in $\mathscr{A}$ which is on the line $u = d$. This is just $\Sigma f(d)g(n)$ summed over those n such that $dn \le x$, or equivalently,

$$f(d) \sum_{n \le x/d} g(n).$$

Summing this last expression over d, $1 \le d \le x$, proves the theorem.

Example. We have already seen the special case of Theorem 22.1 obtained by taking $f = g = \iota_0$, namely

$$\sum_{n \le x} \tau(n) = \sum_{n \le x} \iota_0 \cdot \iota_0(n) = \sum_{d \le x} \iota_0(d) \sum_{n \le x/d} \iota_0(n)$$

$$= \sum_{d \le x} 1 \sum_{n \le x/d} 1 = \sum_{d \le x} [x/d].$$

Example. In Theorem 22.1, take $f = \iota_1$ and $g = \iota_0$. We get

$$\sum_{n \le x} \sigma(n) = \sum_{d \le x} \iota_1(d) \sum_{n \le x/d} \iota_0(n) = \sum_{d \le x} d[x/d]$$

$$= \sum_{d \le x} d(x/d + O(1)) = \sum_{d \le x} x + O\!\left(\sum_{d \le x} d\right)$$

$$= x[x] + O(x^2) = O(x^2).$$

In contrast to the last example, we now show that a sharper result is obtainable by using the commutativity of the product of arithmetic functions.

Theorem 22.2.

$$\sum_{n \le x} \sigma(n) = C_2 x^2 + O(x \log x)$$

for some constant C_2, $1/2 < C_2 < 1$.

Proof. Since $\sigma = \iota_0 \cdot \iota_1$, we have by Theorem 22.1

$$\sum_{n \le x} \sigma(n) = \sum_{d \le x} 1 \sum_{n \le x/d} n$$

$$= \sum_{d \le x} \frac{[x/d]([x/d] + 1)}{2}$$

$$= \frac{1}{2} \sum_{d \le x} \left(\frac{x}{d} + O(1)\right)^2$$

$$= \frac{x^2}{2} \sum_{d \le x} \frac{1}{d^2} + O\left(x \sum_{d \le x} \frac{1}{d}\right) + O\left(\sum_{d \le x} 1\right)$$

$$= \frac{x^2}{2}\left(K - \frac{1}{x} + O(1/x^2)\right) + O(x \log x), \quad \text{from Exercise 20-3,}$$

$$= \frac{Kx^2}{2} - \frac{x}{2} + O(1) + O(x \log x).$$

Replacing $K/2$ by C_2 and noting that $-x/2 + O(1) = O(x \log x)$, we have the theorem proved.

EXERCISES
22-1. Prove that $x \log x = o(x^2) = O(x^2)$. This shows that the error term $O(x \log x)$ in Theorem 22.2 is of smaller order of magnitude than the dominant term $C_2 x^2$.
22-2. Prove that for $k \in \mathscr{Z}^+$,

$$\sum_{n \le x} \sigma_k(n) = O(x^{k+1}).$$

(*Hint*: Use Theorem 22.1 and the result in Exercise 20-5.)

23. Sums Involving the Möbius Function

To motivate the introduction of further machinery, consider an application of Theorem 22.1 to find an asymptotic formula for the function $\Sigma \varphi(n)$. Since $\varphi = \iota_1 \cdot \mu = \mu \cdot \iota_1$,

$$\sum_{n \le x} \varphi(n) = \sum_{n \le x} \iota_1 \cdot \mu(n) = \sum_{d \le x} \iota_1(d) \sum_{n \le x/d} \mu(n)$$

and we cannot proceed without some knowledge of the function $\Sigma_{n \le x} \mu(n)$. By applying the identity $\varphi = \mu \cdot \iota_1$, we get

$$\sum_{n \le x} \varphi(n) = \sum_{d \le x} \mu(d) \sum_{n \le x/d} \iota_1(n) = \sum_{d \le x} \mu(d) \sum_{n \le x/d} n$$

$$= \sum_{d \le x} \mu(d) \frac{[x/d]([x/d] + 1)}{2}$$

(11) $$= \frac{x^2}{2} \sum_{d \le x} \frac{\mu(d)}{d^2} + O\left(x \sum_{d \le x} \frac{\mu(d)}{d}\right) + O\left(\sum_{d \le x} \mu(d)\right).$$

It now appears that we need to know $\Sigma \mu(n) \, n^{-k}$ for $k = 0, 1,$ and 2. Such sums involving the function μ can be evaluated with the use of our next two theorems.

Theorem 23.1. *Suppose f and g are functions of a real variable. If*

(12)
$$f(x) = \sum_{n \leq x} g\left(\frac{x}{n}\right),$$

then

(13)
$$g(x) = \sum_{n \leq x} \mu(n) f\left(\frac{x}{n}\right).$$

Conversely, (13) *implies* (12).

Proof. Suppose (12) holds. Then

$$\sum_{d \leq x} \mu(d) f\left(\frac{x}{d}\right) = \sum_{d \leq x} \mu(d) \sum_{n \leq x/d} g\left(\frac{x/d}{n}\right).$$

As in the proof of Theorem 22.1 we see that this is simply the sum of evaluations of $\mu(d) g(x/nd)$ at each lattice point (n,d) in the region bounded by $u = 1/2$, $v = 1/2$, $uv = x$. Changing the order of summation we evaluate $\mu(a) g(x/ab)$ at each lattice point (a,b) on $uv = n$, $1 \leq n \leq x$, and then sum from $n = 1$ to $n = [x]$. Therefore,

$$\sum_{d \leq x} \mu(d) f\left(\frac{x}{d}\right) = \sum_{n \leq x} \sum_{ab = n} \mu(a) g\left(\frac{x}{ab}\right)$$

$$= \sum_{n \leq x} g\left(\frac{x}{n}\right) \sum_{ab = n} \mu(a)$$

$$= \sum_{n \leq x} g\left(\frac{x}{n}\right) \varepsilon(n) = g(x)$$

since $\varepsilon(n) = 1$ if and only if $n = 1$, zero otherwise. This prove (13).

Conversely, assume (13). Then

$$\sum_{d \leq x} g\left(\frac{x}{d}\right) = \sum_{d \leq x} \sum_{n \leq x/d} \mu(n) f\left(\frac{x}{nd}\right) = \sum_{n \leq x} \sum_{ab = n} \mu(a) f\left(\frac{x}{ab}\right)$$

$$= \sum_{n \leq x} f\left(\frac{x}{n}\right) \varepsilon(n) = f(x).$$

Corollary 23.1.

$$\sum_{n \leq x} \frac{\mu(n)}{n} = O(1)$$

Proof. Take $g(x) = 1$. Then

$$f(x) = \sum_{n \leq x} 1 = x + O(1).$$

By the inversion formula of the theorem, we have

$$1 = \sum_{n \leq x} \mu(n) f\left(\frac{x}{n}\right) = \sum_{n \leq x} \mu(n) \left\{ \frac{x}{n} + O(1) \right\}$$

$$= x \sum_{n \leq x} \frac{\mu(n)}{n} + \sum_{n \leq x} \mu(n) \, O(1).$$

But since $|\mu(n)| \leq 1$,

$$\sum_{n \leq x} \mu(n) \, O(1) = O(x)$$

and

$$1 = x \sum_{n \leq x} \frac{\mu(n)}{n} + O(x)$$

or

$$\sum_{n \leq x} \frac{\mu(n)}{n} = \frac{1}{x} + O(1) = O(1).$$

We now introduce the Dirichlet series, which will be used to study other sums involving the Möbius function. If $f \in \mathscr{A}$, then a series of the type

$$\sum_{n=1}^{\infty} \frac{f(n)}{n^s}, \quad s \text{ real},$$

is called a *Dirichlet series*. The convergence or divergence of these series is determined by the function f and the value of s. It is easy to prove if $f(n) = O(1)$ and $s > 1$, then

$$\sum_{n=1}^{\infty} f(n) n^{-s}$$

is absolutely convergent. If

$$F(s) = \sum_{n=1}^{\infty} f(n) \, n^{-s} \quad \text{and} \quad G(s) = \sum_{n=1}^{\infty} g(n) \, n^{-s}$$

are two Dirichlet series, then the product $F(s) \, G(s)$ is defined by

$$(14) \qquad F(s) \, G(s) = \sum_{n=1}^{\infty} \sum_{m=1}^{\infty} f(n) \, g(m) \, n^{-s} m^{-s}$$

if this series is convergent.

Theorem 23.2. *If $F(s) = \Sigma f(n) \, n^{-s}$ and $G(s) = \Sigma g(n) \, n^{-s}$ are two Dirichlet series which are absolutely convergent, then $F(s) \, G(s)$ is defined and*

$$F(s) \, G(s) = \sum_{n=1}^{\infty} \frac{f \cdot g(n)}{n^s}.$$

Proof. It is known that the product series is absolutely convergent under the given conditions, so the terms in the product series may be rearranged without affecting the sum. Let $n \in \mathscr{Z}^+$ be arbitrary. If $n = ab$ is any factorization of n into positive integers, the series (14) contains a term

$$f(a) g(b) a^{-s} b^{-s} = \frac{f(a) g(b)}{n^s}.$$

Selecting out of (14) all such terms for factorizations of n, we see that (14) contains

$$\sum_{ab=n} \frac{f(a) g(b)}{n^s} = f \cdot g(n) n^{-s}$$

and no other terms with n^{-s} involved. Since n was arbitrary, we have

$$F(s) G(s) = \sum_{n=1}^{\infty} \frac{f \cdot g(n)}{n^s}.$$

Remark. If $s > 1$, the Dirichlet series $\zeta(s) = \Sigma \iota_0(n) n^{-s} = \Sigma 1/n^s$ is the *Riemann zeta function.* It is known that $\zeta(2) = \pi^2/6$ and $\zeta(4) = \pi^4/90$.

Corollary 23.2a.

$$\sum_{n=1}^{\infty} \frac{\mu(n)}{n^s} = \frac{1}{\zeta(s)}, \quad s > 1$$

Proof.

$$\left\{ \Sigma \frac{\mu(n)}{n^s} \right\} \left\{ \Sigma \frac{\iota_0(n)}{n^s} \right\} = \Sigma \frac{\mu \cdot \iota_0(n)}{n^s} = \Sigma \frac{\varepsilon(n)}{n^s}$$

Applying definitions, we have $(\Sigma \mu(n) n^{-s}) \zeta(s) = 1$.

Corollary 23.2b.

$$\sum_{n \leq x} \frac{\mu(n)}{n^2} = \frac{6}{\pi^2} + O\left(\frac{1}{x}\right)$$

Proof.

$$\sum_{n \leq x} \frac{\mu(n)}{n^2} = \sum_{n=1}^{\infty} \frac{\mu(n)}{n^2} - \sum_{n=[x]+1}^{\infty} \frac{\mu(n)}{n^2}$$

$$= \frac{1}{\zeta(2)} + O\left(\sum_{n>x} \frac{1}{n^2}\right)$$

$$= \frac{6}{\pi^2} + O\left(\int_x^{\infty} \frac{dt}{t^2}\right)$$

$$= \frac{6}{\pi^2} + O\left(\frac{1}{x}\right).$$

Putting the appropriate results into (11) we obtain the following.

Theorem 23.3.

$$\sum_{n \leq x} \varphi(n) = \frac{3x^2}{\pi^2} + O(x)$$

Our next theorem is an example of a somewhat different type of application of Theorem 23.2. Before proceeding, the reader should review Theorem 14.3.

Theorem 23.4. *If $s > 1$ and if $c_n(d)$ is Ramanujan's trigonometric function, then*

$$\frac{\sigma_{s-1}(n)}{n^{s-1}} = \zeta(s) \sum_{d=1}^{\infty} \frac{c_n(d)}{d^s}.$$

Proof.

$$\zeta(s) \sum_{d=1}^{\infty} \frac{c_n(d)}{d^s} = \sum_{d=1}^{\infty} \frac{\iota_0 \cdot c_n(d)}{d^s} = \sum_{d|n} \frac{d}{d^s}$$

$$= \frac{1}{n^{s-1}} \sum_{d|n} \left(\frac{n}{d}\right)^{s-1} = \frac{1}{n^{s-1}} \sigma_{s-1}(n)$$

EXERCISES

23-1. Prove that if $f(n) = O(1)$ and $s > 1$, then $\Sigma f(n) n^{-s}$ is absolutely convergent.

23-2. Fill in the details of the proof of Theorem 23.3.

23-3. Find an asymptotic expression for $\Sigma_{n \leq x} \mu(n) n^{-4}$.

23-4. If s and k are suitably restricted, prove
   (a) $\zeta^2(s) = \Sigma \tau(n) n^{-s}$
   (b) $\zeta(s) \zeta(s-1) = \Sigma \sigma(n) n^{-s}$
   (c) $\zeta(s) \zeta(s-k) = \Sigma \sigma_k(n) n^{-s}$
   (d) $\zeta(s-1)/\zeta(s) = \Sigma \varphi(n) n^{-s}$

23-5. Consider again the asymptotic formula of Exercise 20-3. Show that the limit of the left side as $x \to \infty$ is the infinite series $\zeta(2)$, and the limit of the right side is the constant K of Exercise 20-3.

24. Squarefree Integers

We have shown that $\mu^2(n)$ is 1 if n is squarefree, 0 otherwise. Hence $\Sigma_{n \leq x} \mu^2(n)$ is the number of squarefree integers not exceeding x, and we will now study this function. We first notice that

$$\mu^2(n) = \sum_{d^2|n} \mu(d)$$

because n can be written uniquely in the form $n = N^2 q$, q squarefree. Hence,

$$\sum_{d^2|n} \mu(d) = \sum_{d|N} \mu(d) = \varepsilon(N) = \mu^2(n).$$

Theorem 24.1.

$$\sum_{n \leq x} \mu^2(n) = \frac{6x}{\pi^2} + O(\sqrt{x})$$

Proof. In the summations below, we are summing (in two ways) over the lattice points in the region bounded by $u = 1/2$, $v = 1/2$, and $u^2 v = x$.

$$\sum_{n \leq x} \mu^2(n) = \sum_{n \leq x} \sum_{d^2|n} \mu(d)$$

$$= \sum_{d \leq \sqrt{x}} \mu(d) \sum_{n \leq x/d^2} 1$$

$$= \sum_{d \leq \sqrt{x}} \mu(d) \left\{ \frac{x}{d^2} + O(1) \right\}$$

$$= x \sum_{d \leq \sqrt{x}} \frac{\mu(d)}{d^2} + O(\sqrt{x})$$

$$= x\{1/\zeta(2) + O(1/\sqrt{x})\} + O(\sqrt{x})$$

$$= \frac{6x}{\pi^2} + O(\sqrt{x})$$

Chapter 5

SUMS OF SQUARES

25. Sums of Four Squares

In this section we will prove that if $n \in \mathscr{Z}^+$, then there are integers x_j $(j = 1, 2, 3, 4)$ such that $n = x_1^2 + x_2^2 + x_3^2 + x_4^2$; that is, every positive integer can be written as a sum of four squares. The proof is greatly simplified by noting the identity

$$
\sum_{j=1}^{4} x_j^2 \sum_{j=1}^{4} y_j^2 = (x_1y_1 + x_2y_2 + x_3y_3 + x_4y_4)^2
$$
$$
+ (x_1y_2 - x_2y_1 + x_3y_4 - x_4y_3)^2
$$
(1)
$$
+ (x_1y_3 - x_3y_1 + x_4y_2 - x_2y_4)^2
$$
$$
+ (x_1y_4 - x_4y_1 + x_2y_3 - x_3y_2)^2.
$$

Because of (1) it suffices to prove that every prime can be written as a sum of four squares. Since $2 = 1^2 + 1^2 + 0^2 + 0^2$, we may restrict our attention to odd primes.

Lemma 25.1. If p is an odd prime, there exist $a, b \in \mathscr{Z}$ such that $0 \leq a \leq (p-1)/2$, $0 \leq b \leq (p-1)/2$, and $a^2 + b^2 + 1 \equiv 0 \pmod{p}$.

Proof. Consider the sets

$$
\mathscr{A} = \left\{ a^2 : a = 0, 1, \ldots, \frac{p-1}{2} \right\},
$$

$$
\mathscr{B} = \left\{ -b^2 - 1 : b = 0, 1, \ldots, \frac{p-1}{2} \right\}.
$$

Clearly no two a's in $\mathscr{A}$ are congruent $\pmod{p}$, for if $a_1^2 \equiv a_2^2$, then either $a_1 - a_2 \equiv 0$ or $a_1 + a_2 \equiv 0$. If $a_1 \equiv a_2 \pmod{p}$, then $a_1 = a_2$ since $0 \leq a_j \leq (p-1)/2$; if $a_1 + a_2 \equiv 0$, then $a_1 = a_2 = 0$. Similarly, no two elements in $\mathscr{B}$ are congruent. But in $\mathscr{A}$ and $\mathscr{B}$ there are $p + 1$ elements, so that some $a^2 \in \mathscr{A}$ is congruent (mod p) to some $-b^2 - 1 \in \mathscr{B}$. This proves the lemma.

From the lemma we have $a^2 + b^2 + 1^2 + 0^2 = kp$ for some $k > 0$; also, since

$$0 < kp = a^2 + b^2 + 1 \leq 2\left(\frac{p-1}{2}\right)^2 + 1 < p^x,$$

$k < p$. We have the following result:

There exist $x_j \in \mathscr{Z}$ $(j = 1, 2, 3, 4)$ such that

(2)
$$kp = \sum_{j=1}^{4} x_j^2,$$

$0 < k < p$, and $x_j \not\equiv 0 \,(\text{mod } p)$ for at least one x_j.

If we can show that the least k for which (2) holds is $k = 1$, then we will have shown that every odd prime p is expressible as a sum of four squares. Let m be the least positive k for which (2) holds.

Assume $m > 1$. If m is even, then we may assume that

$$x_1 \equiv x_2 \,(\text{mod } 2) \quad \text{and} \quad x_3 \equiv x_4 \,(\text{mod } 2).$$

That is, if m is even, then either none, two, or four of the x_j are even and we may choose notation so that the above congruences hold. Then

$$\left\{\frac{x_1 + x_2}{2}\right\}^2 + \left\{\frac{x_1 - x_2}{2}\right\}^2 + \left\{\frac{x_3 + x_4}{2}\right\}^2 + \left\{\frac{x_3 - x_4}{2}\right\}^2 = \frac{m}{2}p$$

is a representation of $(m/2)p$ as a sum of the squares of four *integers*; by the way m was chosen, from (2) we conclude that

$$\frac{x_1 + x_2}{2} \equiv \frac{x_1 - x_2}{2} \equiv \frac{x_3 + x_4}{2} \equiv \frac{x_3 - x_4}{2} \equiv 0 \,(\text{mod } p).$$

But then

$$\frac{x_1 + x_2}{2} \pm \frac{x_1 - x_2}{2} \equiv 0 \,(\text{mod } p)$$

implies that $x_1 \equiv x_2 \equiv 0$; similarly, $x_3 \equiv x_4 \equiv 0 \,(\text{mod } p)$ and this contradicts (2). Hence m is odd.

Since $m \geq 3$ by assumption, we can find y_j in an absolutely least CRS (mod m), i.e., $|y_j| < m/2$, such that $y_j \equiv x_j \,(\text{mod } m)$ for $j = 1, 2, 3, 4$. Equivalently, there exist m_j such that

(3)
$$y_j = x_j + m_j m, \qquad j = 1, 2, 3, 4.$$

Now

$$\sum_{i=1}^{4} y_j^2 = \sum x_j^2 + 2m \sum x_j m_j + m^2 \sum m_j^2$$

$$\equiv \sum x_j^2 \equiv 0 \,(\text{mod } m).$$

Therefore, for some M we have $Mm = \Sigma y_j^2$. Since $|y_j| < m/2$,

$$0 \le Mm < 4(m/2)^2 = m^2.$$

If $M = 0$, then $y_j = 0$ for all j. But then from (3), $x_j = -m_j m$, and from (2) we get $mp = m^2 \Sigma(-m_j)^2$, or $m|p$, which is impossible. Therefore, $0 < Mm < m^2$, so $0 < M < m$.

We now have $\Sigma x_j^2 = mp$, $\Sigma y_j^2 = Mm$, and $0 < M < m < p$. Let

$$w_1 = x_1 y_1 + x_2 y_2 + x_3 y_3 + x_4 y_4$$

$$w_2 = x_1 y_2 - x_2 y_1 + x_3 y_4 - x_4 y_3$$

$$w_3 = x_1 y_3 - x_3 y_1 + x_4 y_2 - x_2 y_4$$

$$w_4 = x_1 y_4 - x_4 y_1 + x_2 y_3 - x_3 y_2$$

and using (1) we find that

(4)
$$\sum_{j=1}^{4} w_j^2 = Mm^2 p.$$

Notice that $w_1 = \Sigma x_j y_j = \Sigma x_j(x_j + m_j m) = \Sigma x_j^2 + m\Sigma x_j y_j \equiv 0 \,(\text{mod } m)$. In a similar manner, $w_2 \equiv w_3 \equiv w_4 \equiv 0 \,(\text{mod } m)$. We write $w_j = W_j m$ and from (4) get $\Sigma W_j^2 = Mp$. From (2) we conclude $W_j \equiv 0 \,(\text{mod } p)$, $j = 1, 2, 3, 4$. Say $W_j = D_j p$; then $Mp = \Sigma W_j^2 = p^2 \Sigma D_j^2$, from which we have $p|M$, and this is impossible. Hence $m = 1$ and every prime is the sum of four squares.

Together with (1), this proves the following theorem.

Theorem 25.1. *Every positive integer can be written as the sum of four squares.*

Example. To express $231 = 3 \cdot 7 \cdot 11$ as a sum of four squares, we use (1) as follows:

$$3 \cdot 7 = (1^2 + 1^2 + 1^2 + 0^2)(2^2 + 1^2 + 1^2 + 1^2)$$

$$= (2 + 1 + 1 + 0)^2 + (1 - 2 + 1 - 0)^2$$

$$+ (1 - 2 + 0 - 1)^2 + (1 - 0 + 1 - 1)^2$$

$$= 4^2 + 0^2 + (-2)^2 + 1^2.$$

Then again by (1) we have

$$(3 \cdot 7) \cdot 11 = (4^2 + 2^2 + 1^2 + 0^2)(3^2 + 1^2 + 1^2 + 0^2)$$
$$= (12 + 2 + 1 + 0)^2 + (4 - 6 + 0 - 0)^2$$
$$+ (4 - 3 + 0 - 0)^2 + (0 - 0 + 2 - 1)^2$$
$$= 15^2 + (-2)^2 + 1^2 + 1^2.$$

EXERCISES

25-1. Express 105, 385, and 625 as sums of four squares.

25-2. In the example given in this section, we showed that 3 and 11 can be written as a sum of three squares. Prove that 7 *cannot* be expressed as a sum of three squares.

25-3. Prove if $n \equiv -1 \pmod 8$, then n cannot be written as a sum of three squares. [*Hint*: If x, y, z are in a CRS (mod 8), consider the possible values for $x^2 + y^2 + z^2 \pmod 8$.]

26. Sums of Two Squares

If there are integers x and y such that $n = x^2 + y^2$ and $(x,y) = 1$, then we say n has a *primitive representation* as a sum of two squares; if $(x,y) > 1$, we call $n = x^2 + y^2$ an *imprimitive* representation.

It is obvious that not every integer can be written as a sum of two squares. For example, $3 = x^2 + y^2$ has no solution x,y in integers. We will first determine necessary and sufficient conditions for $n \in \mathcal{Z}^+$ to be represented as a sum of two squares. When n can be so represented, we will determine the number of ways in which this can be done.

Lemma 26.1. *Suppose* $n \in \mathcal{Z}^+$. *If there is a prime* q *such that* $q|n$ *and* $q \equiv 3 \pmod 4$, *then* n *has no primitive representation as a sum of two squares.*

Proof. Suppose there exists a prime q such that $q|n$, $q \equiv 3 \pmod 4$, and $n = x^2 + y^2$ is a primitive representation. Evidently, $q \nmid x$ and $q \nmid y$, because if, say, $q|x$, then $q|(n - x^2) = y^2$, so $q|y$, which contradicts our assumption that $(x,y) = 1$.

Since $(x,q) = 1$, we know by Corollary 12.2 that there is some u such that $y \equiv ux \pmod q$. Then

$$x^2(1 + u^2) = x^2 + u^2x^2 \equiv x^2 + y^2 = n \equiv 0 \pmod q,$$

so that $1 + u^2 \equiv 0$, or $u^2 \equiv -1 \pmod q$. Thus $(-1/q) = 1$. But q is of the form $q \equiv 3 \pmod 4$ and by Euler's criterion

$$(-1/q) = (-1)^{(q-1)/2} = -1.$$

Therefore, n has no primitive representation.

Theorem 26.1. *If q is prime, $q \equiv 3 \pmod 4$, $q^\alpha \| n$, α odd, then n has neither primitive nor imprimitive representations as a sum of two squares.*

Proof. Suppose $n = x^2 + y^2$ and $(x,y) = d$. Say $q^\beta \| d$, $\beta \geq 0$. Let $x = Xd$, $y = Yd$, with $(X,Y) = 1$. We have $n = d^2(X^2 + Y^2) = d^2 N$, say. Since $q^{\alpha - 2\beta} \| N$, we have a contradiction because $\alpha - 2\beta$ is odd, hence ≥ 1, and $q|N$, $N = X^2 + Y^2$ is a primitive representation, which is impossible by the preceding lemma.

The problem of writing any n as a sum of two squares may be facilitated by using the identity

(5) $$(x_1^2 + x_2^2)(y_1^2 + y_2^2) = (x_1 y_1 + x_2 y_2)^2 + (x_1 y_2 - x_2 y_1)^2,$$

which tells us that if n_1 and n_2 are each representable as a sum of two squares, then so is $n_1 n_2$. If n is any positive integer, suppose the canonical form is

$$n = 2^\alpha \prod p_i^{\alpha_i} \prod q_j^{\beta_j}$$

where $p_i \equiv 1 \pmod 4$ and $q_j \equiv 3 \pmod 4$. We know from Theorem 26.1 that β_j must be even for all j if n is to have a representation; indeed, if β_j is even, then $q_j^{\beta_j} = \{q_j^{(1/2)\beta_j}\}^2 + 0^2$ is a representation of $q_j^{\beta_j}$, and by using (5) we can write $\prod q_j^{\beta_j}$ as a sum of two squares when the β_j are even. Also $2 = 1^2 + 1^2$ together with (5) shows that every power of 2 is a sum of two squares. Now we will prove that if $p \equiv 1 \pmod 4$, then $p = x^2 + y^2$ has a solution $x,y \in \mathscr{Z}$. For this, we require the following lemma.

Lemma 26.2. *If z is real and n is a positive integer, then there are integers a and b such that*

$$\left| z - \frac{a}{b} \right| \leq \frac{1}{b(n + 1)}, \qquad 1 \leq b \leq n.$$

Proof. Consider the numbers $\alpha_k = kz - [kz]$, $k = 0, 1, \ldots, n$. Obviously, $0 \leq \alpha_k < 1$ for $k = 0, 1, \ldots, n$. By renaming, if necessary, call these $n + 1$ numbers β_k such that $0 \leq \beta_0 \leq \beta_1 \leq \cdots \leq \beta_n$, and consider the differences

$$\beta_1 - \beta_0, \beta_2 - \beta_1, \ldots, \beta_n - \beta_{n-1}, \beta_0 - \beta_n + 1.$$

Each of these $n + 1$ numbers is non-negative, and

$$(\beta_1 - \beta_0) + (\beta_2 - \beta_1) + \cdots + (\beta_0 - \beta_n + 1) = 1;$$

therefore, at least one of them is $\leq 1/(n + 1)$. But each of these, and in particular the one which is $\leq 1/(n + 1)$, has the form $\beta_k - \beta_j + r$, where $k \neq j, 0 \leq j \leq n, 0 \leq k \leq n$, and r is 0 or 1. Hence

$$\frac{1}{n + 1} \geq \beta_k - \beta_j + r = (k - j)z + [jz] - [kz] + r \geq 0.$$

If $k > j$, take $b = k - j$ and $a = [kz] - [jz] - r$, and if $k < j$, take $b = j - k$ and $a = [jz] - [kz] + r$. In either case, $|bz - a| \leq 1/(n + 1)$. Clearly, since j,k are distinct non-negative integers not exceeding n, we have $1 \leq b = |k - j| \leq n$, and $|z - a/b| \leq 1/b(n + 1)$.

Now if p is a prime, $p \equiv 1 \pmod 4$, we use Lemma 26.2 to show that p is the sum of two squares. Since -1 is a quadratic residue of p, there is some u such that $u^2 \equiv -1 \pmod p$. In the above lemma, let $z = -u/p$ and $n = [\sqrt p]$. There are $a,b \in \mathscr{Z}$ such that

$$\left| -\frac{u}{p} - \frac{a}{b} \right| \leq \frac{1}{b([\sqrt p] + 1)} < \frac{1}{b\sqrt p}, \qquad 0 < b \leq [\sqrt p] \leq \sqrt p.$$

Let $c = ub + ap$, so that

$$|c| = | - (ub + ap)| = \left| -\frac{u}{p} - \frac{a}{b} \right| |pb| < \sqrt p.$$

Therefore, $0 < b^2 + c^2 < (\sqrt p)^2 + (\sqrt p)^2 = 2p$. But $c \equiv ub \pmod p$, and $b^2 + c^2 \equiv b^2(1 + u^2) \equiv 0 \pmod p$. Since $0 < b^2 + c^2 < 2p$, we must have $p = b^2 + c^2$.

Together with our previous results this proves Theorem 26.2.

Theorem 26.2. *Suppose*

$$n = 2^\alpha \prod_{i=1}^r p_i^{\alpha_i} \prod_{j=1}^s q_j^{\beta_j}, \qquad \alpha \geq 0, \qquad \alpha_i > 0, \qquad \beta_j > 0,$$

$p_i \equiv 1 \pmod 4$, $q_j \equiv 3 \pmod 4$. *The equation $n = x^2 + y^2$ has an integral solution x,y if and only if β_j is even for $j = 1,\ldots,s$.*

Example. For $n = 2^3 5 \cdot 13 \cdot 11^2$, we note that $2^3 = 2^2 + 2^2$, $5 = 2^2 + 1^2$, $13 = 2^2 + 3^2$, and $11^2 = 11^2 + 0^2$. Now by successive applications of (5) we have

$$2^3 5 = (2^2 + 2^2)(2^2 + 1^2) = (4 + 2)^2 + (2 - 4)^2 = 6^2 + 2^2,$$

$$(2^3 5)13 = (6^2 + 2^2)(2^2 + 3^2) = (12 + 6)^2 + (18 - 4)^2 = 18^2 + 14^2,$$

$$(2^3 5 \cdot 13)11^2 = (18^2 + 14^2)(11^2 + 0^2) = (198 + 0)^2 + (0 - 154)^2.$$

EXERCISES

26-1.* Show that $2^{2\alpha} = (\pm 2^\alpha)^2 + 0^2$ and $2^{2\alpha+1} = (\pm 2^\alpha)^2 + (\pm 2^\alpha)^2$ are the only representations as sums of two squares for powers of 2.

26-2. Express 45, 325, and 5929 as a sum of two squares.

26-3. Prove that for every $\beta \in \mathscr{Z}^+$, the equation $3^{2\beta} = x^2 + y^2$ has only the four solutions $x = \pm 3^\beta$, $y = 0$ and $x = 0$, $y = \pm 3^\beta$.

27. Number of Representations

Exercise 26-1 shows that a power of 2 may be written as a sum of two squares in four ways, namely, $2^{2\alpha} = (2^\alpha)^2 + 0^2 = (-2^\alpha)^2 + 0^2 = 0^2 + (2^\alpha)^2 = 0^2 + (-2^\alpha)^2$. This is a special case of the general problem which we now consider: If $n \in \mathscr{L}^+$, in how many ways can n be written as a sum of two squares? Suppose

$$n = 2^\alpha \prod_{k=1}^{r} p_k^{\beta_k} \prod_{j=1}^{s} q_j^{\gamma_j}$$

where $\alpha \geq 0$, $p_k \equiv 1 \pmod 4$, $q_j \equiv 3 \pmod 4$. For convenience we write $n_1 = \Pi p_k^{\beta_k}$ and $n_2 = \Pi q_j^{\gamma_j}$, so we have $n = 2^\alpha n_1 n_2$.

Theorem 27.1. *If n is as above, then the number $N(n)$ of ways in which n can be written as a sum of two squares is*

$$N(n) = \begin{cases} 4\tau(n_1) & \text{if } n_2 \text{ is a square}; \\ 0 & \text{if } n_2 \text{ is not a square}. \end{cases}$$

The proof of this theorem will be postponed until we investigate some of the properties of the Gaussian integers in the next section.

28. The Gaussian Integers

A *Gaussian integer* is a complex number of the form $a + bi$, where $a, b \in \mathscr{L}$ and $i^2 = -1$. Let $\mathscr{G}$ denote the set of all Gaussian integers. Two Gaussian integers $a + bi$ and $c + di$ are *equal* if and only if $a = c$ and $b = d$. Addition, subtraction, and multiplication in $\mathscr{G}$ are defined just as these operations are defined on the set of all complex numbers, namely

$$(a + bi) \pm (c + di) = (a \pm c) + (b \pm d)i$$

$$(a + bi)(c + di) = (ac - bd) + (ad + bc)i.$$

Evidently, if $\alpha, \beta \in \mathscr{G}$, then $\alpha \pm \beta$ and $\alpha\beta \in \mathscr{G}$. Also, the set $\mathscr{L}$ of rational integers is a subset of $\mathscr{G}$, and the multiplicative identity in $\mathscr{G}$ is the (Gaussian) integer 1. A Gaussian integer α is called a *unit* in $\mathscr{G}$ if and only if there exists $\beta \in \mathscr{G}$ such that $\alpha\beta = 1$.

A mapping D, called the *norm*, from $\mathscr{G}$ into the set of non-negative rational integers is defined by $D(a + bi) = a^2 + b^2$. This function has the property that $D(\alpha\beta) = D(\alpha) D(\beta)$ for all $\alpha, \beta \in \mathscr{G}$ (the proof is left as an exercise). Thus it is easy to see that α is a unit in $\mathscr{G}$ only if $D(\alpha)|D(1) = 1$, hence only if $D(\alpha) = 1$, since $D(\alpha) \geq 0$. But if $\alpha = a + bi$, $D(\alpha) = a^2 + b^2 = 1$ implies either $a = \pm 1$

and $b = 0$, or $a = 0$ and $b = \pm 1$. Therefore, if α is a unit, α must be of the form ± 1 or $\pm i$. Clearly, these four numbers are units and hence the only units in $\mathcal{G}$. We notice that α is a unit if and only if $D(\alpha) = 1$, and the units are all of the form i^n, $n \in \mathcal{Z}$.

Two integers $\alpha, \beta \in \mathcal{G}$ are called *associates* if $\alpha = \beta \gamma$ for some *unit* γ. Thus α and β are associates if and only if $\alpha = \pm \beta$ or $\alpha = \pm \beta i$. If $\alpha, \beta \in \mathcal{G}$, we say α *divides* β, and write $\alpha | \beta$, if and only if there exists $\gamma \in \mathcal{G}$ such that $\alpha \gamma = \beta$. Clearly a unit is a divisor of every Gaussian integer. A non-zero, non-unit element $\alpha \in \mathcal{G}$ is called a *prime* if the only divisors of α are units or associates of α.

If $\alpha \in \mathcal{G}$ and $D(\alpha)$ is a rational prime, then α is prime in $\mathcal{G}$, for if $\alpha = \beta \gamma$ is any factorization, then $D(\alpha) = D(\beta) D(\gamma)$ and since $D(\alpha)$ is a prime in $\mathcal{Z}$, either $D(\beta)$ or $D(\gamma) = 1$. We will show later that there are primes $\delta \in \mathcal{G}$ such that $D(\delta) \in \mathcal{Z}$ is composite.

Example. The Gaussian integers $1 + i$, $2 + i$, $1 - i$, $2 - i$ are primes in $\mathcal{G}$ since their norms are prime in $\mathcal{Z}$. The primes $1 + i$ and $1 - i$ are associated, since $(1 + i)(-i) = 1 - i$.

Every element of $\mathcal{G}$ which is not zero and not a unit can be factored into primes, as we now show. Let $\alpha \neq 0$, α not a unit. If α is a prime, we are through. If α is not a prime, then α has a representation of the form $\alpha = \beta \gamma$ with $D(\beta) > 1$ and $D(\gamma) > 1$. Evidently we also have $D(\beta) < D(\alpha)$ and $D(\gamma) < D(\alpha)$ since $D(\alpha) = D(\beta) D(\gamma)$. Now if β and γ are not primes, we may continue the factorization process. Since the norms of the factors form a strictly decreasing sequence of positive integers, the process must terminate when every further factorization involves a factor whose norm is 1.

Thus, a non-zero, non-unit in $\mathcal{G}$ may be written as a product of primes in $\mathcal{G}$. To show that this factorization is essentially (except for order and the occurrence of units and associates) unique, we require the Gaussian analogue of the division algorithm for rational integers.

Theorem 28.1. *If* $\alpha, \beta \in \mathcal{G}$, $\beta \neq 0$, *then there exist* $\lambda, \theta \in \mathcal{G}$ *such that* $\alpha = \beta \lambda + \theta$, $D(\theta) < D(\beta)$.

Proof. The proof is greatly facilitated by operating in the set of all complex numbers, rather than in the set $\mathcal{G}$. Recall that if γ is a complex number, γ has a unique representation of the form $\gamma = x + yi$ with x, y real. Also, the *conjugate* $\bar{\gamma}$ of γ is the complex number $\bar{\gamma} = x - yi$. Equality, addition (and subtraction), and multiplication of complex numbers are defined just as these concepts are defined in $\mathcal{G}$, and if $\gamma = x + yi$, $\delta = u + vi$, $\delta \neq 0$ are complex numbers, then division γ/δ is defined by

$$\frac{\gamma}{\delta} = \frac{xu + yv}{u^2 + v^2} + \frac{yu - xv}{u^2 + v^2} i.$$

Finally, the *modulus* $|\gamma|$ of a complex number γ is defined by $|\gamma| = (\gamma\bar{\gamma})^{1/2}$. In particular, notice that if the complex number α is also a Gaussian integer, then $D(\alpha) = |\alpha|^2 = \alpha\bar{\alpha}$.

Now suppose $\alpha,\beta \in \mathscr{G}$, $\beta \neq 0$. Then α/β is defined and is some complex number, say $\alpha/\beta = x + yi$ where x and y are rational. We now choose rational integers m and n such that $|x - m| \leq 1/2$, $|y - n| \leq 1/2$. This can be done by using Lemma 26.2 with the n of that lemma equal to 1.

Take $\lambda = m + ni$ and $x = \alpha - \beta\lambda$. We must show that $D(x) < D(\beta)$. But we have

$$|x| = |\alpha - \beta\lambda| = |\beta|\left|\frac{\alpha}{\beta} - \lambda\right| = |\beta|\,|(x - m) + (y - n)i|$$

$$= |\beta|\{(x - m)^2 + (y - n)^2\}^{1/2}$$

$$\leq |\beta|\{(\tfrac{1}{2})^2 + (\tfrac{1}{2})^2\}^{1/2} < |\beta|.$$

Hence $D(x) = |x|^2 < |\beta|^2 = D(\beta)$.

With this theorem we can define a Euclidean algorithm in $\mathscr{G}$ as follows. Suppose $\alpha,\beta \in \mathscr{G}$, $\beta \neq 0$. We find integers in $\mathscr{G}$ such that

$$\alpha = \beta\lambda_1 + x_1, \qquad D(x_1) < D(\beta)$$

$$\beta = x_1\lambda_2 + x_2, \qquad D(x_2) < D(x_1)$$

$$\cdots$$

$$x_{n-2} = x_{n-1}\lambda_n + x_n, \qquad D(x_n) < D(x_{n-1})$$

$$x_{n-1} = x_n\lambda_{n+1} + x_{n+1}, \qquad D(x_{n+1}) = 0.$$

The process must terminate with some remainder x_{n+1} such that $D(x_{n+1}) = 0$, since the norms form a strictly decreasing sequence of non-negative rational integers. The remainder x_n is a common divisor of α and β, and every common divisor divides x_n. Hence, we call x_n a *greatest common divisor* of α and β and write $x_n = (\alpha,\beta)$. This gcd is not unique, since any associate of x_n is also a gcd. However, any two gcd's must be associates, for suppose $\gamma = (\alpha,\beta)$ and $\delta = (\alpha,\beta)$. Then $\gamma|\delta$, say $\gamma\sigma = \delta$, and $\delta|\gamma$, say $\delta\pi = \gamma$. Therefore $\gamma = \delta\pi = \gamma\sigma\pi$, so $\sigma\pi = 1$, σ is a unit and γ and δ are associates.

The Euclidean algorithm allows us to prove that if $(\alpha,\beta) = 1$ and $\alpha|\beta\gamma$, then $\alpha|\gamma$. In addition, then, if α is a prime in $\mathscr{G}$ and $\alpha|\beta\gamma$, either $\alpha|\beta$ or $\alpha|\gamma$. This in turn yields a proof of the essential uniqueness of factorization in $\mathscr{G}$. The details are very similar to the proof of the same theorem for rational integers, and are omitted here.

Theorem 28.2. *If $\alpha \in \mathcal{G}$ is a non-zero, non-unit, then α can be written as a product of primes in $\mathcal{G}$, and the factorization is essentially unique.*

We consider now the problem of determining the primes in $\mathcal{G}$. We have already shown that $1 + i, 1 - i, -1 + i$, and $-1 - i$ are all prime in $\mathcal{G}$ since the norms are 2, and evidently these primes are associates.

Suppose q is a rational prime and $q \equiv 3 \pmod 4$. Let $g = \alpha\beta$ be any factorization of q in $\mathcal{G}$. We then have $D(q) = q^2 = D(\alpha) D(\beta)$. By unique factorization of rational integers, either $D(\alpha) = 1, q$, or q^2. If $D(\alpha) = 1$, then α is a unit; also, if $D(\alpha) = q^2$, then β is a unit. If $D(\alpha) = D(\beta) = q$, and if $\alpha = a + bi$, then $D(\alpha) = q = a^2 + b^2$. But q has no representation as a sum of two squares. Therefore, every factorization of $q \equiv 3 \pmod 4$ involves a unit factor, so q is prime in $\mathcal{G}$. *Note that $D(q)$ is not a rational prime.*

Suppose p is a rational prime, $p \equiv 1 \pmod 4$. We recall that p can be written as a sum of squares, say $p = a^2 + b^2$. Thus, in $\mathcal{G}$ we have the factorization $p = (a + bi)(a - bi)$. Notice that $a + bi$ and $a - bi$ are not associates, for if they were, we would have $(a + bi)i^n = a - bi$ for some n, $0 \le n \le 3$. But $ai^n + bi^{n+1} = a - bi$ is impossible since a and b are relatively prime and non-zero. Also, the numbers $a \pm bi$ are primes in $\mathcal{G}$, since $D(a \pm bi) = p$.

The primes determined above are in fact the only primes in $\mathcal{G}$, as we will show.

Theorem 28.3. *The primes in $\mathcal{G}$ are precisely the numbers $1 + i$, the rational primes $q \equiv 3 \pmod 4$, and the numbers $a \pm bi$, where $a^2 + b^2$ is a rational prime $p \equiv 1 \pmod 4$, and associates of these.*

Proof. We have already shown that each of the numbers described in the theorem is prime in $\mathcal{G}$. To complete the proof that there are no others, let α be a prime of $\mathcal{G}$. Since $D(\alpha) = \alpha\bar{\alpha}$, $\alpha | D(\alpha)$. Thus, there are rational positive integers which are divisible by α. Let m be the least positive integer such that $\alpha | m$. If m is not a rational prime, then $m = n_1 n_2$, with $1 < n_j < m$ ($j = 1,2$), and $\alpha | n_1$ or $\alpha | n_2$ contrary to the way m was chosen. Thus α divides at least one rational prime. Suppose p and q are distinct rational primes such that $\alpha | p$ and $\alpha | q$. Since $(p,q) = 1$, there are $a, b \in \mathcal{Z}$ such that $pa + qb = 1$, and $\alpha | (pa + qb) = 1$, which is impossible since α is prime. Therefore, a prime $\alpha \in \mathcal{G}$ divides exactly one positive rational prime.

Let $\alpha = a + bi$ and let p be the prime in $\mathcal{Z}$ such that $\alpha\beta = p$ for some $\beta \in \mathcal{G}$. Taking norms, we have $D(\alpha) | p^2$. Hence, either $D(\alpha) = p$, or $D(\alpha) = p^2$ in which case $D(\beta) = 1$ and α is an associate of p.

Case 1, $p = 2$.
If $D(\alpha) = 2 = a^2 + b^2$, we have $a = \pm 1$, $b = \pm 1$. If $D(\alpha) = 4$, then α is an associate of $2 = (1 + i)(1 - i)$, which is not prime in $\mathcal{G}$.

Case 2, p ≡ 1 (mod 4).

If $D(\alpha) = p$, then $\alpha = \pm a \pm bi$, where $p = a^2 + b^2$; α cannot be an associate of p, because we have seen p is not prime in $\mathscr{G}$.

Case 3, p ≡ 3 (mod 4).

We know $D(\alpha) = p$ is impossible. Thus, α is an associate of p.

EXERCISES

28-1.   Find $(1 - 3i, 3 + 5i)$; express $1 - 3i$ and $3 + 5i$ as products of primes in $\mathscr{G}$.

28-2.   Prove that if $\alpha, \beta \in \mathscr{G}$, then $D(\alpha\beta) = D(\alpha) D(\beta)$. Show that $D(\alpha) = 0$ if and only if $\alpha = 0$. Show that $D(\bar{\alpha}) = D(\alpha)$, and that $D(n\alpha) = n^2 D(\alpha)$ for all $n \in \mathscr{Z}$.

28-3.   Prove or disprove: For all $\alpha, \beta \in \mathscr{G}$, $D(\alpha) + D(\beta) \le D(\alpha + \beta)$.

28-4.   Suppose p is a prime in $\mathscr{Z}$. Prove that the Legendre symbol $(-4/p)$ is

(a)   undefined if and only if p is the product of two associated conjugate primes in $\mathscr{G}$ (then p is said to *ramify* in $\mathscr{G}$);

(b) $= -1$ if and only if p stays prime in $\mathscr{G}$;

(c) $= +1$ if and only if p is the product of two non-associated conjugate primes in $\mathscr{G}$ (p *splits* in $\mathscr{G}$).

28-5.   Give an example to show that for some $\alpha, \beta \in \mathscr{G}$, $\beta \ne 0$, there exist $\lambda_1, \theta_1, \lambda_2, \theta_2 \in \mathscr{G}$ such that

$$\alpha = \beta\lambda_1 + \theta_1, \qquad D(\theta_1) < D(\beta)$$

$$\alpha = \beta\lambda_2 + \theta_2, \qquad D(\theta_2) < D(\beta)$$

and $D(\theta_1) \ne D(\theta_2)$.

28-6.   Show that if $(\alpha, \beta) = 1$ and $\alpha | \beta\gamma$, then $\alpha | \gamma$. Use this to supply the details of the proof of Theorem 28.2.

28-7.   If $\alpha, \beta \in \mathscr{G}$, $\alpha\beta \ne 0$, consider the complex numbers μ satisfying

$$\mu = \frac{\alpha\beta}{(\alpha, \beta)}, \quad \text{for some } (\alpha, \beta).$$

Prove that

(a)   if μ satisfies the above equation, then $\mu \in \mathscr{G}$;

(b)   if μ_1 and μ_2 satisfy the above equation, then μ_1 and μ_2 are associates;

(c) $\alpha | \mu$ and $\beta | \mu$;

(d)   if $\alpha | \gamma$ and $\beta | \gamma$, then $\mu | \gamma$.

28-8.   Suppose $a, b \in \mathscr{Z}$ and not both a and b are zero. If $(a, b)_{\mathscr{G}}$ denotes their gcd in $\mathscr{G}$ and $(a, b)_{\mathscr{Z}}$ denotes their gcd in $\mathscr{Z}$, prove that $(a, b)_{\mathscr{G}} = \pm(a, b)_{\mathscr{Z}}$.

29. Proof of Theorem 27.1

We are now able to prove Theorem 27.1. Suppose $n = 2^\alpha n_1 n_2$ where $n_1 = \Pi p^\beta$, $n_2 = \Pi q^\gamma$, and $p \equiv 1$ (mod 4), $q \equiv 3$ (mod 4). By Theorem 26.2, the number $N(n)$ of representations of n as a sum of two squares is 0 if n_2 is not a square, so we may assume that each γ is even, say $\gamma = 2\delta$. If $n = x^2 + y^2$, then in $\mathscr{G}$, $n = (x + yi)(x - yi)$, and we may count the number of pairs x,y by considering the factorizations of $x \pm yi$ in $\mathscr{G}$. We have

$$n = i^\mu (1 + i)^\alpha (1 - i)^\alpha \prod_{p|n} (a + bi)^\beta (a - bi)^\beta \prod_{q|n} q^{2\delta}$$

where the factor i^μ ($\mu = 0, 1, 2, 3$) allows all possible associated factors and $p = a^2 + b^2 = (a + bi)(a - bi)$ for each $p \equiv 1$ (mod 4). If $n = (x + yi)(x - yi)$, then

(6) $\qquad (x + yi) = i^{M'}(1 + i)^{A'}(1 - i)^{A''}\Pi(a + bi)^{B'}(a - bi)^{B''}\Pi q^{D'}$

and

(7) $\qquad (x - yi) = i^{M_1}(1 + i)^{A_1}(1 - i)^{A_2}\Pi(a + bi)^{B_1}(a - bi)^{B_2}\Pi q^{D_1}$

where $M' + M_1 = \mu$, $A' + A_1 = \alpha = A'' + A_2$, $B' + B_1 = \beta = B'' + B_2$, and $D' + D_1 = 2\delta$.

But we can in fact say much more about these exponents. It is easy to show that if π and σ are complex numbers, then $\overline{\pi\sigma} = \bar{\pi}\bar{\sigma}$. Hence (7), as the conjugate of (6), is

(7') $\qquad x - yi = (-i)^{M'}(1 - i)^{A'}(1 + i)^{A''}\Pi(a - bi)^{B'}(a + bi)^{B''}\Pi q^{D'}.$

By comparing (7) and (7') we see that $A'' = A_1$, $A' = A_2$, $B'' = B_1$, $B' = B_2$, $D' = D_1$. With our previous observations about these exponents, we conclude that

(8) $\qquad x + yi = i^{M'}(1 + i)^{A'}(1 - i)^{\alpha - A'}\Pi(a + bi)^{B'}(a - bi)^{\beta - B'}\Pi q^{D'}$

(9) $\qquad x - yi = i^{\mu - M'}(1 + i)^{\alpha - A'}(1 + i)^{A'}\Pi(a + bi)^{\beta - B'}(a - bi)^{B'}\Pi q^{D'}.$

Thus, different factorizations may be obtained by choosing $0 \le M' \le 3$, $0 \le A' \le \alpha, 0 \le B' \le \beta$. However, notice that

$$(1 + i)^{A' + 1}(1 - i)^{\alpha - (A' + 1)} = (1 + i)^{A'}(1 - i)^{\alpha - A'}(1 + i)(1 - i)^{-1}$$

$$= (1 + i)^{A'}(1 - i)^{\alpha - A'}i.$$

Hence, all the variations obtained by choosing different values for A' may be obtained by choosing different M'. That this is not the case with the factors $a \pm bi$ is seen as follows. Suppose for some choice of M, M', B, B' we have

$$i^M(a + bi)^B(a - bi)^{\beta - B} = i^{M'}(a + bi)^{B'}(a - bi)^{\beta - B'}.$$

By uniqueness of factorization in $\mathscr{G}$, either $B = B'$, or $(a + bi)^B$ is an associate of $(a - bi)^{\beta - B'}$. But we have already seen that $a + bi$ and $a - bi$ are non-associates.

Therefore, distinct expressions in (8) and (9) are found by taking $0 \le M' \le 3$, and $0 \le B' \le \beta$. Then $N(n) = 4\Pi(\beta + 1) = 4\tau(n_1)$, as was to be proved.

30. Restatement of Theorem 27.1

Theorem 27.1 is sometimes stated in another form. We define the following arithmetic functions. Let

$$\tau_k(n) = \sum_{\substack{d|n \\ d \equiv k \,(\text{mod } 4)}} 1 \quad \text{for } k = 1 \text{ and } 3.$$

Thus, $\tau_1(n)$ is the number of divisors of n which are of the form $4m + 1$, and $\tau_3(n)$ is the number of divisors of the form $4m + 3$.

Theorem 30.1.

$$N(n) = 4\{\tau_1(n) - \tau_3(n)\}$$

The proof of this theorem requires some knowledge of the functions τ_1 and τ_3. Notice that neither of these functions is multiplicative; for example, $\tau_1(3) = \tau_1(7) = \tau_3(3) = \tau_3(7) = 1$, but $\tau_1(21) = 2 = \tau_3(21)$. Although not multiplicative, the functions do have these interesting properties: If $(a,b) = 1$, then

(10) $$\tau_1(ab) = \tau_1(a)\,\tau_1(b) + \tau_3(a)\,\tau_3(b)$$

(11) $$\tau_3(ab) = \tau_1(a)\,\tau_3(b) + \tau_1(b)\,\tau_3(a).$$

Equation (10) holds because every divisor d of ab may be written uniquely as $d = AB$, where $A|a$, $B|b$; then $d \equiv 1 \pmod 4$ if and only if $A \equiv B \equiv 1$ (mod 4) or $A \equiv B \equiv 3$ (mod 4), and $d \equiv 3 \pmod 4$ if and only if $A \equiv 1$, $B \equiv 3$ or $A \equiv 3$, $B \equiv 1 \pmod 4$. But there are $\tau_1(ab)$ ways to choose $d \equiv 1$, while there are $\tau_1(a)\,\tau_1(b)$ ways to choose $A \equiv B \equiv 1$ plus $\tau_3(a)\,\tau_3(b)$ ways to choose $A \equiv B \equiv 3 \pmod 4$, and this proves (10). In a similar manner (11) is proved. The remainder of the proof of Theorem 30.1 is easy to complete (see Exercises below).

EXERCISES

30-1. Suppose $n = 2^x n_1 n_2$, where the canonical form of n_1 contains only primes $p \equiv 1 \pmod 4$ and n_2 contains only primes $q \equiv 3 \pmod 4$. Use (10) and (11) to prove that $\tau_k(n) = \tau_k(n_1 n_2)$ for $k = 1, 3$.

30-2. Define the arithmetic function F by $F(n) = \tau_1(n) - \tau_3(n)$. Use Exercise 30-1 to prove that F is multiplicative.

30-3. With n as in Exercise 30-1, and F as in Exercise 30-2, prove $F(2^z) = 1$, $F(n_1) = \tau(n_1)$, and $F(n_2) = 1$ or 0 according as n_2 is a square or not. Then use Exercise 30-2 to conclude that $F(n)$ is $\tau(n_1)$ if n_2 is a square, 0 otherwise. (*Hint:* Use Theorem 8.1.)

Chapter 6

CONTINUED FRACTIONS, FAREY SEQUENCES, THE PELL EQUATION

IN THIS CHAPTER we will study continued fractions, Farey sequences, and the Pell equation. Each of these terms will be defined later, but roughly speaking, continued fractions are certain numbers written in the form

$$x_0 + \cfrac{1}{x_1 + \cfrac{}{\ddots + \cfrac{1}{x_n}}}$$

where the x_i are appropriately restricted real numbers; Farey sequences are certain ordered collections of fractions between 0 and 1; and the Pell equation is the Diophantine equation $x^2 - Dy^2 = 1$, $D \in \mathscr{Z}$. We will observe some interrelations of the three topics, and we conclude the chapter with some results concerning the approximation of real numbers by rationals.

31. Finite Continued Fractions

Definition. *Let $x_0, x_1, \ldots, x_n$ be real numbers with $x_i \neq 0$ ($i = 1, \ldots, n$). A finite continued fraction with j entries ($j = 0, 1, \ldots, n$) will be denoted $\|x_0, \ldots, x_j\|$ and is defined inductively by*

$$\|x_0\| = x_0,$$

$$\|x_0, x_1\| = x_0 + x_1^{-1},$$

$$\|x_0, x_1, \ldots, x_j\| = \|x_0, \|x_1, \ldots, x_j\| \|, \quad 1 < j \leq n.$$

98

If $x_0 \in \mathscr{L}$ and $x_i \in \mathscr{L}^+$ $(i = 1, \ldots, n)$, the continued fraction $\|x_0, x_1, \ldots, x_n\|$ will be called simple. The number x_j $(j = 0, 1, \ldots, n)$ is called the jth partial quotient of the continued fraction.

Examples.

(a) $$\|1,2,3\| = 1 + \|2,3\|^{-1} = 1 + (2 + 3^{-1})^{-1} = \tfrac{10}{7}$$

(b) $$\|-2,1,2,3\| = -2 + \|1,2,3\|^{-1} = \frac{-13}{10}$$

$$= -2 + \|1 + \|2,3\|^{-1}\|^{-1}$$

$$= \|-2,1 + \|2,3\|^{-1}\|$$

$$= \|-2,1, \|2,3\|\,\|$$

(c) $$\|-2,2,1\| = -2 + \|2,1\|^{-1}$$

$$= -2 + 3^{-1} = \|-2,3\|$$

Example (b) above illustrates the following theorem for $j = 1$, $n = 3$.

Theorem 31.1.

$$\|x_0, x_1, \ldots, x_n\| = \|x_0, \ldots, x_j, \|x_{j+1}, \ldots, x_n\|\,\|, \quad 0 \le j < n, \quad n \ge 1$$

Proof. For $n = 1$ the theorem is obviously true for all j, $0 \le j < n = 1$. Suppose the theorem has been proved up to and including $n = N - 1 \ge 1$ (this supposition is referred to below as the primary inductive assumption). We will now show the theorem is true for N (thus completing the induction on n) by using induction on j. With $j = 0$, the result is immediate from the definition of a continued fraction. Suppose the result has been established for j, $0 \le j < t < N$ (this is the secondary inductive assumption). Then

$$\|x_0, \ldots, x_N\| = \|x_0, \ldots, x_{t-1}, \|x_t, \ldots, x_N\|\,\|, \quad \text{by the secondary assumption with } j = t - 1,$$

$$= \left\|x_0, \ldots, x_{t-1}, \|x_t, \|x_{t+1}, \ldots, x_N\|\,\|\,\right\|, \quad \text{from the definition,}$$

$$= \|x_0, \ldots, x_{t-1}, x_t, \|x_{t+1}, \ldots, x_N\|\,\|, \quad \text{by the primary assumption with } n = t < N.$$

The proof is complete.

An interesting special case of Theorem 31.1 occurs as follows.

Corollary 31.1. If $x_n \ne 1$, $n \ge 1$, then

$$\|x_0, \ldots, x_{n-1}, x_n - 1, 1\| = \|x_0, \ldots, x_{n-1}, x_n\|.$$

Proof. Since $\|x_n - 1,1\| = x_n$,

$$\|x_0,\dots,x_{n-1},x_n - 1,1\| = \|x_0,\dots,x_{n-1},\|x_n - 1,1\|\,\|$$
$$= \|x_0,\dots,x_{n-1},x_n\|.$$

From the corollary it is clear that any number which has a finite continued fraction representation has at least two such representations, as illustrated in Example (c). If a continued fraction is not simple, it is evident that there is an unlimited number of non-trivial possible changes in the x_j which do not change the value of the continued fraction. However, if the fraction is simple, then there is only the one change of the type in Corollary 31.1 which does not change the value. The situation is described more completely in Theorem 31.2.

Theorem 31.2. *If $\|a_0,\dots,a_n\|$ is simple, then it has a unique rational value. If r is any rational number, then there are exactly two finite simple continued fractions with value r, and exactly one of these continued fractions has its last partial quotient equal to 1.*

Proof. The first statement of the theorem is easily proved by induction on the number of partial quotients. If every simple continued fraction with n partial quotients is rational, and if $\|a_0,\dots,a_n\|$ is any continued fraction with $n + 1$ partial quotients, then $\|a_0,\dots,a_n\| = \|a_0,\|a_1,\dots,a_n\|\,\|$ is rational.

We will now show that if r is rational, then r can be represented by a finite simple continued fraction, and that if $\|a_0,\dots,a_m\| = \|b_0,\dots,b_n\|$ are two equal simple continued fractions with $a_m > 1$ and $b_n > 1$, then $m = n$ and $a_i = b_i$ for $i = 0,1,\dots,n$.

If r is rational, define the sequence of integers $a_0, a_1, \dots, a_n$ as follows. Take $a_0 = [r]$, the largest integer not exceeding r. Then

$$r = a_0 + \theta_0, \qquad 0 \le \theta_0 < 1.$$

If $\theta_0 = 0$, we are through; if $\theta_0 > 0$, then $1/\theta_0 > 1$ and we take $a_1 = [1/\theta_0] \ge 1$. We have

$$\frac{1}{\theta_0} = a_1 + \theta_1, \qquad 0 \le \theta_1 < 1.$$

Continue by induction. If $\theta_j \ne 0$, define $\theta_{j+1} = 1/\theta_j - [1/\theta_j]$ and take $a_{j+1} = [1/\theta_j]$; if $\theta_j = 0$, stop the process with a_j and $j = n$. This process (called the *continued fraction algorithm*) must stop after a finite number of steps, because $\theta_0 = r - a_0$ is rational. By induction, then, θ_j is the difference of two rationals, hence is rational for every j. Say

$$\theta_j = \frac{c_j}{d_j}, \qquad 0 \le c_j < d_j, \qquad (c_j, d_j) = 1.$$

Suppose $c_j \neq 0$. Then

$$\frac{d_j}{c_j} = \frac{1}{\theta_j} = a_{j+1} + \frac{c_{j+1}}{d_{j+1}}, \qquad 0 \le c_{j+1} < d_{j+1}$$

gives

$$\frac{c_j c_{j+1}}{d_{j+1}} = d_j - c_j a_{j+1} \in \mathscr{Z}.$$

Therefore, $d_{j+1} | c_j c_{j+1}$, but $(d_{j+1}, c_{j+1}) = 1$ implies that $d_{j+1} | c_j$, so $d_{j+1} \le c_j$ and hence $0 \le c_{j+1} < c_j$ for all j. Thus, the c_j's form a decreasing sequence of non-negative integers so for some j, $c_j = 0$.

We now show that the simple fraction $\|a_0, \ldots, a_n\|$, with a_j defined as above, has the value r. We have

$$r = a_0 + \theta_0 = \|a_0, \theta_0^{-1}\| = \|a_0, a_1 + \theta_1\|$$

$$= \|a_0, \|a_1, \theta_1^{-1}\|\| = \|a_0, a_1, \theta_1^{-1}\| = \cdots = \|a_0, \ldots, a_{n-1}, \theta_{n-1}^{-1}\|$$

$$= \|a_0, \ldots, a_{n-1}, a_n + \theta_n\|$$

but $\theta_n = 0$.

Finally, suppose

(1) $$\|a_0, \ldots, a_m\| = \|b_0, \ldots, b_n\|, \qquad a_m > 1, \quad b_n > 1.$$

Then

(2) $$a_0 = [\|a_0, \ldots, a_m\|] = [\|b_0, \ldots, b_n\|] = b_0.$$

Since $a_0 = b_0$, it follows from (1) and the definition of a continued fraction that $\|a_1, \ldots, a_m\| = \|b_1, \ldots, b_n\|$. If $m \le n$, we apply the same argument $m - 1$ times to find $a_j = b_j$ $(j = 1, \ldots, m - 1)$, and we have $\|a_m\| = \|b_m, \ldots, b_n\|$. But $\|a_m\|$ is an integer, and $\|b_m, \ldots, b_n\|$ can be an integer only if $m = n$, since $b_n > 1$. Hence $m = n$ and $a_m = b_m$.

Definition. *Let* $\|a_0, a_1, \ldots, a_n\|$ *be a finite continued fraction. For* $0 \le k \le n$, *the* k^{th} *convergent* C_k *to the given continued fraction is the continued fraction* $\|a_0, \ldots, a_k\|$.

We will now define two sequences $\{P_k\}$ and $\{Q_k\}$ associated with a continued fraction such that $C_k = P_k/Q_k$, $k = 0, 1, \ldots, n$. For a given continued fraction $\|a_0, a_1, \ldots, a_n\|$, we define

(3)
$$P_0 = a_0, \qquad P_1 = a_1 a_0 + 1, \qquad P_k = a_k P_{k-1} + P_{k-2} \ (k \geq 2),$$
$$Q_0 = 1, \qquad Q_1 = a_1, \qquad \qquad Q_k = a_k Q_{k-1} + Q_{k-2} \ (k \geq 2).$$

Obviously if $\|a_0, \ldots, a_n\|$ is simple, then $P_k \in \mathscr{Z}$, $Q_k \in \mathscr{Z}^+$ ($k = 0, 1, \ldots, n$), and $Q_k \geq k$. In Exercise 31-1, we outline a proof of the claim that $C_k = P_k/Q_k$, $k = 0, 1, \ldots, n$.

Theorem 31.3. *For $0 \leq k < n$, define*

$$\Delta(k) = P_k Q_{k+1} - P_{k+1} Q_k.$$

Then $\Delta(k) = (-1)^{k+1}$.

Proof. For $k = 0$ and $k = 1$, the theorem is immediate from (3). If $\Delta(k-1) = (-1)^k$ for $k - 1 \geq 1$, then

$$P_k Q_{k+1} - P_{k+1} Q_k = P_k(a_{k+1} Q_k + Q_{k-1}) - (a_{k+1} P_k + P_{k-1})Q_k$$
$$= -\Delta(k-1) = (-1)^{k+1}$$

Corollary 31.3a. *If $\|a_0, a_1, \ldots, a_n\|$ is simple, the convergents C_k are rational and in lowest terms.*

Proof. The theorem leads to a solution of the Diophantine equation $P_k x + Q_k y = 1$, so $(P_k, Q_k) = 1$.

Corollary 31.3b.

$$C_{k+1} - C_k = \frac{(-1)^k}{Q_k Q_{k+1}}$$

Proof. This corollary is proved by dividing through the equation $\Delta(k) = (-1)^{k+1}$ by $-Q_k Q_{k+1}$.

We will now prove that the convergents oscillate about the value of the continued fraction with decreasing amplitude.

Theorem 31.4. *Suppose $\alpha = \|a_0, a_1, \ldots, a_n\|$. Then*

$$C_0 < C_2 < C_4 < \cdots \leq \alpha = C_n \leq \cdots < C_5 < C_3 < C_1.$$

Proof. We have

$$C_{2k+2} - C_{2k} = \frac{P_{2k+2}}{Q_{2k+2}} - \frac{P_{2k}}{Q_{2k}}$$
$$= \frac{-a_{2k+2} \Delta(2k)}{Q_{2k} Q_{2k+2}} > 0$$

so the convergents with even subscripts form an increasing sequence. Also

$$C_{2k+1} - C_{2k-1} = \frac{-a_{2k+1}\Delta(2k-1)}{Q_{2k-1}Q_{2k+1}} < 0$$

so the convergents with odd subscripts form a decreasing sequence.

Finally we must show that any convergent C_{2k} is less than every convergent C_{2j+1}. For $2k < 2j + 1$, Corollary 31.3b gives $C_{2j+1} - C_{2j} > 0$, so $C_{2j+1} > C_{2j} \geq C_{2k}$, since $2k \leq 2j$ and the even convergents are increasing. For $2j + 1 < 2k$, Corollary 31.3b gives $C_{2k+1} - C_{2k} > 0$, so $C_{2k+1} > C_{2k} \geq C_{2j}$.

EXERCISES

31-1. For $k = 0$ and $k = 1$, verify that

(4) $$C_k = \frac{P_k}{Q_k}.$$

Assume that (4) has been proved for the kth convergent ($k \geq 1$) of every continued fraction with at least $k + 1$ entries, and prove (4) holds for $k + 1$ by noting that $C_{k+1} = \|a_0,\ldots,a_k,a_{k+1}\| = \|a_0,\ldots,a_k + a_{k+1}^{-1}\|$ is the kth convergent of some continued fraction.

31-2. Consider again the numbers c_j/d_j defined in the proof of Theorem 31.2. If $r = a/b$, $(a,b) = 1$, prove that $b = d_0$, $d_j = c_{j-1}$ ($j = 1,\ldots,n$), and that $d_n = c_{n-1} = 1$.

31-3. With c_j ($j = 0,1,\ldots,n$) as in Exercise 31-2, prove that c_{j+1} is the least non-negative residue of c_{j-1} (mod c_j) for $j = 1,2,\ldots,n-1$.

31-4. Suppose r is rational, say $r = a/b$, $(a,b) = 1$. Use the Euclidean algorithm to find

$$a = ba_0 + c_0, \qquad 0 < c_0 < b$$
$$b = c_0a_1 + c_1, \qquad 0 < c_1 < c_0$$
$$\cdots$$
$$c_{n-2} = c_{n-1}a_n + c_n, \qquad c_n = 0.$$

Prove that these a_j are the a_j defined in the proof of Theorem 31.2, and these c_j/c_{j+1} are the c_j/d_j of Theorem 31.2.

32. Infinite Simple Continued Fractions

We now extend the concept of a continued fraction to an *infinite* simple continued fraction.

Definition. *Let* $a_1, a_2, a_3, \ldots$ *be any sequence of natural numbers. The* infinite simple continued fraction $\|a_0,a_1,a_2,\ldots\|$, $a_0 \in \mathscr{Z}$, *is defined by*

$$\|a_0,a_1,a_2,\ldots\| = \lim_{n\to\infty} \|a_0,a_1,\ldots,a_n\|$$

if this limit exists. If the limit exists and is equal to ρ, then we say $\|a_0,a_1,a_2,\ldots\|$ has value ρ, or represents the real number ρ.

The question of the existence of the above limit is easily answered in the affirmative. In fact, the limit not only exists, but is always irrational.

Theorem 32.1. *The infinite simple continued fraction $\|a_0,a_1,\ldots\|$ is always defined; that is, if $a_0 \in \mathscr{Z}$ and $a_n \in \mathscr{Z}^+$ $(n = 1,2,\ldots)$, then*

$$\lim_{n \to \infty} \|a_0,a_1,\ldots,a_n\|$$

exists and is irrational. If ρ is any irrational number, then ρ has a representation as an infinite simple continued fraction, and the representation is unique.

Proof. As before, we let

$$\|a_0,a_1,\ldots,a_n\| = C_n = \frac{P_n}{Q_n}$$

denote the nth convergent. Since the proof of the oscillatory character of the convergents given in Theorem 31.4 did not depend upon the finiteness of the continued fraction, we also know that for infinite continued fractions the even convergents are monotone increasing, the odd convergents are monotone decreasing, and any even convergent is less than every odd convergent. Therefore, $\{C_{2k}\}$ is increasing and bounded above by C_1, and so converges, say to U. Similarly, $\{C_{2k+1}\}$ is decreasing and bounded below by C_0, and so converges, say to L. Clearly L and U must be the same, because if ε is any positive number, then L and U differ by less than ε. Choose $n \in \mathscr{Z}^+$ sufficiently large so that $1/4n^2 < \varepsilon$; then notice that $C_{2n} \le L \le C_{2n+1}$ (since C_{2n} is a lower bound for $\{C_{2k+1}\}$ and L is the greatest lower bound of $\{C_{2k+1}\}$) and $C_{2n} \le U \le C_{2n+1}$ (since C_{2n+1} is an upper bound for $\{C_{2n}\}$ and U is the least upper bound); hence

$$|L - U| \le |C_{2n+1} - C_{2n}| = \frac{(-1)^{2n}}{Q_{2n}Q_{2n+1}} \le \frac{1}{2n(2n + 1)} < \frac{1}{4n^2} < \varepsilon.$$

(We have used here our previous remark that $Q_k \ge k$ for simple continued fractions.) Thus we see that the sequence $\{C_n\}$ of convergents converges to $L\,(=U)$.

To show that L is irrational, we first observe that 0 cannot be represented by an infinite simple continued fraction, because if $0 = \|x_0,x_1,\ldots\|$, for example, then since the convergents oscillate about the value of the continued fraction we must have

$$x_0 < 0 < x_0 + x_1^{-1}.$$

But $x_0 < 0$ and $x_1 \in \mathscr{Z}^+$, so $x_1^{-1} \leq 1$ and $x_0 + x_1^{-1} \leq 0$; therefore 0 has no infinite simple continued fraction representation. Consequently L above is irrational, for suppose the contrary. Then $L = \|a_0, a_1, \ldots\|$ is rational, and L also has a finite continued fraction representation by Theorem 31.2, say $L = \|b_0, b_1, \ldots, b_N\|$. By taking integral parts of $\|b_0, \ldots, b_N\| = \|a_0, a_1, \ldots\|$ and arguing as in the proof of Theorem 31.2, we get $b_j = a_j$ ($j = 0, 1, \ldots, N$) and $0 = \|a_{N+1}, a_{N+2}, \ldots\|$, which is impossible. Hence every infinite simple fraction has an irrational value.

Conversely, suppose ρ is irrational. That ρ has at least one representation as an infinite simple continued fraction is easily seen using the continued fraction algorithm. We find

$$\rho = a_0 + \theta_0, \qquad 0 \leq \theta_0 < 1 \qquad (a_0 = [\rho])$$

$$\frac{1}{\theta_0} = a_1 + \theta_1, \qquad 0 \leq \theta_1 < 1 \qquad (a_1 = [1/\theta_0])$$

$$\cdots$$

$$\frac{1}{\theta_n} = a_{n+1} + \theta_{n+1}, \qquad 0 \leq \theta_{n+1} < 1 \qquad (a_{n+1} = [1/\theta_n])$$

$$\cdots$$

Since ρ is irrational, $\theta_0 = \rho - a_0$ is irrational and so is $1/\theta_0$. Hence $\theta_1 = 1/\theta_0 - a_1$ is irrational. Suppose we have shown that θ_k is irrational; then $\theta_{k+1} = 1/\theta_k - a_{k+1}$ is irrational, so $\theta_{k+1} \neq 0$. Therefore, the process does not terminate and determines an infinite sequence of integers $a_0, a_1, a_2, \ldots$. Now we show that $\rho = \|a_0, a_1, \ldots\|$. We have $\rho = a_0 + \theta_0 = \|a_0, \theta_0^{-1}\| = \|a_0, a_1 + \theta_1\| = \cdots = \|a_0, a_1, \ldots, a_n, \theta_n^{-1}\|$ for all n. Let

$$C_n = \|a_0, \ldots, a_n\| = \frac{P_n}{Q_n} \quad \text{and} \quad C'_{n+1} = \|a_0, \ldots, a_n, \theta_n^{-1}\| = \frac{P'_{n+1}}{Q'_{n+1}}.$$

Then for any fixed n, the jth convergent to C'_{n+1} is the jth convergent to C_n if $0 \leq j \leq n$; hence $P'_j = P_j$ and $Q'_j = Q_j$ if $0 \leq j \leq n$. Therefore,

$$\rho = \frac{P'_{n+1}}{Q'_{n+1}} = \frac{\theta_n^{-1} P'_n + P'_{n-1}}{\theta_n^{-1} Q'_n + Q'_{n-1}} = \frac{\theta_n^{-1} P_n + P_{n-1}}{\theta_n^{-1} Q_n + Q_{n-1}},$$

and

$$|\rho - C_n| = \left| \rho - \frac{P_n}{Q_n} \right| = \left| \frac{\Delta(n-1)}{(\theta_n^{-1} Q_n + Q_{n-1}) Q_n} \right|$$

$$= \frac{1}{(\theta_n^{-1} Q_n + Q_{n-1}) Q_n} < \frac{1}{(a_{n+1} Q_n + Q_{n-1}) Q_n} = \frac{1}{Q_{n+1} Q_n} \leq \frac{1}{n^2}.$$

Since $1/n^2 \to 0$ as $n \to \infty$, we see that $\lim_{n \to \infty} C_n = \rho$, or that $\|a_0, a_1, \ldots\| = \rho$.
The uniqueness of this representation is proved by arguments similar to those used in proving the analogous result for finite simple continued fractions.

The following corollary is listed here for emphasis.

Corollary 32.1.

$$|\rho - C_n| = \left| \rho - \frac{P_n}{Q_n} \right| < \frac{1}{Q_n Q_{n+1}} < \frac{1}{Q_n^2}$$

for all $n \in \mathscr{L}^+$.

In Chapter 9 we will prove that ρ has a *periodic* infinite simple continued fraction representation if and only if ρ is a quadratic irrational, that is, ρ is an irrational root of an equation $ax^2 + bx + c = 0$, with $a, b, c \in \mathscr{L}$.

EXERCISES

32-1. Prove that the infinite simple continued fraction representation of an irrational number is unique.

32-2. Show that Lemma 26.2 is implied by Corollary 32.1 and that there are infinitely many integers x, y with the properties stated in the lemma.

32-3. Consider the continued fraction $\xi = \|1,1,1,\ldots\|$. Since $\xi = 1 + \|1,1,1,\ldots\|^{-1} = 1 + 1/\xi$, prove that $\xi = (1 + \sqrt{5})/2$.

32-4. With ξ as in Exercise 32-3, prove that $P_n = Q_{n+1}, n = 0, 1, \ldots$ and that

$$Q_n = \frac{\left(\dfrac{1 + \sqrt{5}}{2}\right)^{n+1} - \left(\dfrac{1 - \sqrt{5}}{2}\right)^{n+1}}{\sqrt{5}}, \quad n = 0, 1, 2, \ldots.$$

This sequence $\{Q_n\}$ is called the *Fibonacci sequence.*

32-5.* With ξ as in Exercise 32-3, let $\xi = 1 + \theta_0$, and $\theta_n^{-1} = 1 + \theta_{n+1}$ ($n = 0, 1, 2, \ldots$). Prove that $\theta_n^{-1} = \xi$ for all $n \geq 0$.

32-6.* With ξ as in Exercise 32-3, prove that

$$\lim_{n \to \infty} \frac{Q_n}{Q_{n+1}} = \frac{1}{\xi} = \frac{\sqrt{5} - 1}{2}.$$

With θ_n as in Exercise 32-5, prove that

$$\lim_{n \to \infty} \left(\theta_{n+1}^{-1} + \frac{Q_n}{Q_{n+1}} \right) = \sqrt{5}.$$

33. Farey Sequences

Definition. *Suppose* $m \in \mathscr{Z}^+$ *and let*

$$\mathscr{E}_m = \left\{ \frac{r}{s} : s \neq 0,\ r,s \in \mathscr{Z},\ 0 \leq r \leq s \leq m,\ (r,s) = 1 \right\}.$$

The sequence obtained by arranging the elements of $\mathscr{E}_m$ in increasing order is called the Farey sequence of order m, *and will be denoted by* $\mathscr{F}_m$.
Examples.

$$\mathscr{F}_1 : \frac{0}{1}, \frac{1}{1}$$

$$\mathscr{F}_2 : \frac{0}{1}, \frac{1}{2}, \frac{1}{1}$$

$$\mathscr{F}_3 : \frac{0}{1}, \frac{1}{3}, \frac{1}{2}, \frac{2}{3}, \frac{1}{1}$$

$$\mathscr{F}_4 : \frac{0}{1}, \frac{1}{4}, \frac{1}{3}, \frac{1}{2}, \frac{2}{3}, \frac{3}{4}, \frac{1}{1}$$

The number of terms in $\mathscr{F}_m$ is $v\mathscr{E}_m = \Sigma_{d \leq m}\ \varphi(d) + 1$. If $m < n$, then $\mathscr{E}_m \subset \mathscr{E}_n$. If $a/b < c/d$ are in $\mathscr{E}_m$ and no other element of $\mathscr{E}_m$ is between these two, then we say $a/b < c/d$ are *successive* in $\mathscr{F}_m$, or that c/d is the *successor* of a/b in $\mathscr{F}_m$. The number $(a + c)/(b + d)$ is called the *medIant* of a/b and c/d; we shall see that the mediant plays an interesting role in the theory of Farey sequences.

In our next theorem we give a method for finding the successor of a given term $a/b < 1$ in $\mathscr{F}_m$.

Theorem 33.1. *Suppose* $a/b \in \mathscr{E}_m$, $a < b$. *If* $u = c$, $v = d$ *is the unique solution of* $bu - av = 1$ *satisfying* $m - b < d \leq m$, *then* $a/b < c/d$ *are successive in* $\mathscr{F}_m$.

Proof. Since $(a,b) = 1$, there is a solution u_0, v_0 of the Diophantine equation $bu - av = 1$. Choose $t_0 \in \mathscr{Z}$ to be

$$t_0 = \left[\frac{m - v_0}{b} \right].$$

Since $u = u_0 + at$, $v = v_0 + bt$ is a solution for all $t \in \mathscr{Z}$, it is obvious that there is one and only one solution u,v which satisfies $m - b < v \leq m$. This solution is $c = u_0 + at_0$, $d = v_0 + bt_0$, because

$$\frac{m - v_0}{b} - 1 < t_0 \leq \frac{m - v_0}{b}$$

gives

$$0 \leq m - b < v_0 + bt_0 = d \leq m.$$

Furthermore, $c = (1 + ad)/b > 0$ and

$$c = \frac{1 + ad}{b} \le \frac{1 + am}{b} < \frac{1}{b} + m \le m + 1$$

so $0 < c \le m$. Since c,d is a solution of the given equation, $(c,d) = 1$ and

$$c = \frac{1 + ad}{b} < \frac{1}{b} + d$$

so $c \le d$; hence $c/d \in \mathcal{E}_m$.

Now

$$\frac{c}{d} = \frac{bc}{bd} = \frac{1 + ad}{bd} = \frac{1}{bd} + \frac{a}{b} > \frac{a}{b}$$

so c/d occurs after a/b in $\mathcal{F}_m$. Suppose $a/b < c/d$ are not successive in $\mathcal{F}_m$. Say there is a term $x/y \in \mathcal{E}_m$ such that

$$\frac{a}{b} < \frac{x}{y} < \frac{c}{d}.$$

Then

$$\frac{c}{d} - \frac{x}{y} = \frac{cy - dx}{dy} \ge \frac{1}{dy}$$

because $cy - dx \in \mathcal{Z}^+$, since $0 < c/d - x/y$; and

$$\frac{x}{y} - \frac{a}{b} = \frac{bx - ay}{by} \ge \frac{1}{by}.$$

Therefore,

$$\frac{1}{bd} = \frac{bc - ad}{bd} = \frac{c}{d} - \frac{a}{b} = \left(\frac{c}{d} - \frac{x}{y}\right) + \left(\frac{x}{y} - \frac{a}{b}\right) \ge \frac{1}{dy} + \frac{1}{by}$$

$$= \frac{b + d}{bdy} > \frac{m}{bdy} \ge \frac{1}{bd},$$

which is a contradiction. Hence c/d is the successor of a/b.

Corollary 33.1a. *If $a/b < c/d$ are successive in $\mathcal{F}_m$, then $bc - ad = 1$.*

Corollary 33.1b. *Assume $a/b < u/v$ are successive in $\mathcal{F}_m$ and $u/v < c/d$ are successive in $\mathcal{F}_m$. Then u/v is the mediant $(a + c)/(b + d)$ of a/b and c/d.*
Proof. From Corollary 33.1a we have $bu - av = 1$ and $-du + cv = 1$.

Solving these equations for u,v we get

$$u = \frac{a + c}{bc - ad}, \qquad v = \frac{b + d}{bc - ad},$$

so $u/v = (a + c)/(b + d)$.

We now make some remarks about the mediant $(a + c)/(b + d)$ of two terms $a/b < c/d$ which are successive in $\mathscr{F}_m$. From Corollary 33.1a it is clear that the mediant is a rational number which is between a/b and c/d, i.e.,

$$\frac{a}{b} < \frac{a + c}{b + d} < \frac{c}{d}.$$

Also, the mediant is in lowest terms; that is, $(a + c, b + d) = 1$ because the equation $(a + c)x + (b + d)y = 1$ has a solution $x = -d$, $y = c$. Consequently, $b + d > m$, otherwise the mediant is in $\mathscr{E}_m$ and $a/b < c/d$ are not successive in $\mathscr{F}_m$. These remarks suggest the following theorem.

Theorem 33.2. *Suppose $a/b < c/d$ are successive in $\mathscr{F}_m$. If $b + d = m + 1$, then the mediant $(a + c)/(b + d)$ is the only term between a/b and c/d in $\mathscr{F}_{m+1}$; conversely, if the mediant is between a/b and c/d in $\mathscr{F}_{m+1}$, then the mediant is the only term between them in $\mathscr{F}_{m+1}$ and $b + d = m + 1$. Hence, if $a/b < c/d$ are successive in $\mathscr{F}_m$, either they are also successive in $\mathscr{F}_{m+1}$ or they have exactly one term, their mediant, between them in $\mathscr{F}_{m+1}$, and the latter is the case if and only if $b + d = m + 1$.*

Proof. Assume $a/b < c/d$ are successive in $\mathscr{F}_m$. If $b + d = m + 1$, then the mediant is in $\mathscr{E}_{m+1}$. By Theorem 33.1, $u = a + c$, $v = b + d$ is the unique solution of

$$bu - av = 1, \qquad m + 1 - b < b + d \leq m + 1$$

so $a/b < (a + c)/(b + d)$ are successive in $\mathscr{F}_{m+1}$. Again by Theorem 33.1, $u = c$, $v = d$ is the unique solution of

$$(b + d)u - (a + c)v = 1, \qquad m + 1 - (b + d) < d \leq m + 1$$

so $(a + c)/(b + d) < c/d$ are successive in $\mathscr{F}_{m+1}$. Therefore, the mediant is the only term in $\mathscr{F}_{m+1}$ between a/b and c/d.

Now suppose the mediant is between a/b and c/d in $\mathscr{F}_{m+1}$. Since $b + d > m$ from the remarks above, we have $b + d \geq m + 1$. But the mediant is in $\mathscr{E}_{m+1}$, so $b + d \leq m + 1$, and the condition $b + d = m + 1$ is necessary. Moreover, by the sufficiency of the condition, the mediant is the only term between a/b and c/d in $\mathscr{F}_{m+1}$. This completes the proof.

Theorem 33.1 defines a recursive construction for finding the next term in $\mathscr{F}_m$ if any term less than 1 is given. Theorem 33.2 defines a recursive construction for finding $\mathscr{F}_{m+1}$ from $\mathscr{F}_m$.

EXERCISES

33-1.   Construct $\mathscr{F}_5$ and $\mathscr{F}_6$.

33-2.   Find the successor c/d of 3/17 in $\mathscr{F}_{17}$. Now find the term a/b such that $a/b < 3/17$ are successive in $\mathscr{F}_{17}$.

33-3.   Prove that $a/b \in \mathscr{E}_m$ if and only if $(b - a)/b \in \mathscr{E}_m$ and that $a/b \neq (b - a)/b$ except when $a = 1$, $b = 2$. Use these observations to prove that

$$\sum_{x \in \mathscr{E}_m} x = \tfrac{1}{2} + \tfrac{1}{2} \sum_{d \leq m} \varphi(d) = \tfrac{1}{2}\mathbf{v}\mathscr{E}_m.$$

33-4.* Suppose x', x'' are real numbers satisfying $0 \leq x' < x'' \leq 1$. Prove there is some $m \in \mathscr{Z}^+$ such that $\mathscr{E}_m$ contains an element e satisfying $x' < e < x''$.

34. The Pell Equation

We consider now the Diophantine equation $x^2 - Dy^2 = 1$, $D \in \mathscr{Z}$, which is called the Pell equation. If $D < 0$, then $x^2 - Dy^2 \geq 1$ for all $x, y \in \mathscr{Z}$ (except $x = y = 0$, which is not a solution) and the equality can hold only for $x = \pm 1$, $y = 0$, or for $x = 0$, $y = \pm 1$ if $D = -1$. If D is the square of an integer, say $D = d^2$, let $u = dy$ and the solutions must satisfy

$$(x - u)(x + u) = x^2 - Dy^2 = 1.$$

Hence $x + u = x - u = \pm 1$, and $u = 0$, $x = \pm 1$ are the only solutions. Therefore we see that the only interesting cases of the Pell equation arise when D is positive and not a square, hence if and only if $\sqrt{D}$ is irrational. Consequently, throughout this section we shall assume $\sqrt{D}$ is irrational.

We will say that the solution x, y is a *positive* solution of the Pell equation if and only if both x and y are positive. Clearly, all solutions will be known when all positive solutions are known. Let $\mathscr{S}$ be the set of numbers of the form $x + \sqrt{D}y$ for positive solutions x, y of the Pell equation, i.e.,

$$\mathscr{S} = \{x + \sqrt{D}y : x, y \in \mathscr{Z}^+,\ x^2 - Dy^2 = 1\}.$$

If $\mathscr{S} \neq \varnothing$, then $\mathscr{S}$ has a smallest element, because it happens that any two elements of $\mathscr{S}$ differ by at least 1 (see Exercise 34-1), so if $s \in \mathscr{S}$, the number of elements in $\mathscr{S}$ less than s cannot exceed the number of positive integers less than s. Thus, the elements of $\mathscr{S}$ not larger than s are finitely many, and from these we can select a minimum with no difficulty. Consequently, if the Pell equation has a positive solution, then it has a positive solution x, y for which $x + \sqrt{D}y$ is minimal; the corresponding solution x, y is called the *fundamental solution,* and the minimal sum $x + \sqrt{D}y$ will be called the *fundamental factor.*

We will show that indeed there is always a positive solution (hence a fundamental solution), hence infinitely many positive solutions x,y, and that all the corresponding factors $x + \sqrt{D}y$ are positive integral powers of the fundamental factor. If x,y is a positive solution, we will call $x + \sqrt{D}y$ *the dominant factor* corresponding to the solution x,y, and we will refer to the conjugate $x - \sqrt{D}y$ as the *inferior factor* corresponding to x,y. This terminology is suggested by the following facts: If x,y is a positive solution, then $1 < x + \sqrt{D}y$; consequently, from $(x + \sqrt{D}y)(x - \sqrt{D}y) = 1$ we see that $x - \sqrt{D}y = (x + \sqrt{D}y)^{-1}$ is positive and less than 1, hence

$$0 < x - \sqrt{D}y < 1 < x + \sqrt{D}y.$$

Theorem 34.1. *There exists a positive solution of the Pell equation* $x^2 - Dy^2 = 1$; *hence, there exists a fundamental solution.*

Proof. By Corollary 32.1 there are infinitely many pairs $x,y \in \mathscr{Z}^+$ such that $|\sqrt{D} - x/y| < 1/y^2$, or such that $|x - \sqrt{D}y| = |\sqrt{D}y - x| < 1/y$. Since

$$x - \sqrt{D}y = |x| - |\sqrt{D}y| \le |x - \sqrt{D}y| \le \frac{1}{y} \le y$$

we have $x < (1 + \sqrt{D})y$ and $x + \sqrt{D}y < (1 + 2\sqrt{D})y$. Then for every such pair x,y,

$$|x^2 - Dy^2| = |x - \sqrt{D}y|\,|x + \sqrt{D}y| < \frac{1}{y}(1 + 2\sqrt{D})y = 1 + 2\sqrt{D}.$$

Thus, all of the forms $x^2 - Dy^2$ have integer values between $-(1 + 2\sqrt{D})$ and $1 + 2\sqrt{D}$, so there is an integer k, $-(1 + 2\sqrt{D}) < k < 1 + 2\sqrt{D}$, such that for *infinitely many* x,y we have $x^2 - Dy^2 = k$. Notice $k \ne 0$ since D is not a square, and if $k = 1$ we are through.

If $k \ne 1$, we partition the set of all such pairs x,y into equivalence classes by putting two pairs u,v and x,y into the same class if and only if

(5) $$u \equiv x \,(\mathrm{mod}\,|k|) \quad \text{and} \quad v \equiv y \,(\mathrm{mod}\,|k|).$$

Since there are infinitely many pairs and only k^2 classes, some class contains two pairs x_1, y_1 and x_2, y_2 of positive integers such that $x_1 \ne x_2$ and $y_1 \ne y_2$. Define

$$x_0 = \frac{x_1 x_2 - Dy_1 y_2}{k} \quad \text{and} \quad y_0 = \frac{x_1 y_2 - x_2 y_1}{k}.$$

Since x_1, y_1 and x_2, y_2 satisfy (5), we see that the numerators of x_0 and y_0 are $\equiv 0 \,(\mathrm{mod}\,|k|)$, so $x_0 \in \mathscr{Z}$ and $y_0 \in \mathscr{Z}$. Also,

$$x_0^2 - Dy_0^2 = \frac{1}{k^2}(x_1^2 - Dy_1^2)(x_2^2 - Dy_2^2) = 1.$$

Finally, neither x_0 nor y_0 is zero, because if $x_0 = 0$, then $1 = x_0^2 - Dy_0^2 = -Dy_0^2 \leq 0$. If $y_0 = 0$, then $x_1 = x_2 y_1/y_2$ and

$$k = x_1^2 - Dy_1^2 = \left(\frac{x_2 y_1}{y_2}\right)^2 - Dy_1^2 = \left(\frac{y_1}{y_2}\right)^2 (x_2^2 - Dy_2^2) = \left(\frac{y_1}{y_2}\right)^2 k$$

so $y_1^2 = y_2^2$; but since y_1 and y_2 are positive, we have $y_1 = y_2$, and this contradicts the way the pairs x_1, y_1 and x_2, y_2 were chosen.

We will now observe that the product of dominant factors corresponding to two positive solutions is again a dominant factor corresponding to a positive solution, or what is the same thing, the product of two elements of $\mathscr{S}$ is in $\mathscr{S}$. (The same is true for inferior factors; the proof is left as an exercise.) More precisely, suppose x,y and u,v are positive solutions of the Pell equation. Then the pair a,b of integers defined by

$$(x + \sqrt{D}y)(u + \sqrt{D}v) = a + \sqrt{D}b$$

is a positive solution. Clearly, $a = xu + Dyv \in \mathscr{Z}^+$ and $b = xv + yu \in \mathscr{Z}^+$. Also,

$$a^2 - Db^2 = (x^2 - Dy^2)(u^2 - Dv^2) = 1,$$

which proves the remark. An obvious corollary is that if x_1, y_1 is the fundamental solution, then for every $n \in \mathscr{Z}^+$, $(x_1 + \sqrt{D}y_1)^n$ is the dominant factor of a positive solution, so there are infinitely many solutions. The remarkable thing is that the dominant factor of *every* positive solution has this form, which we prove next.

Theorem 34.2. *The set $\mathscr{S}$ of dominant factors corresponding to the positive solutions of the Pell equation is precisely the set of positive integral powers of the dominant fundamental factor.*

Proof. Let x_1, y_1 be the fundamental solution. We have already proved that $\{(x_1 + \sqrt{D}y_1)^n : n \in \mathscr{Z}^+\} \subset \mathscr{S}$, so all that remains to be shown is that if u,v is a positive solution, then there exists $n \in \mathscr{Z}^+$ such that

$$u + \sqrt{D}v = (x_1 + \sqrt{D}y_1)^n.$$

Suppose $u + \sqrt{D}v$ is not a power of $x_1 + \sqrt{D}y_1$. Let $\alpha = x_1 + \sqrt{D}y_1$; then α^{-1} is the inferior fundamental factor. Because of the supposition, we can find $n \in \mathscr{Z}^+$ such that

$$\alpha^n < u + \sqrt{D}v < \alpha^{n+1}$$

so that

(6) $$1 = \alpha^n \alpha^{-n} < (u + \sqrt{D}v)\alpha^{-n} < \alpha^{n+1}\alpha^{-n} = \alpha.$$

Since the product of inferior factors is an inferior factor, $\alpha^{-n} = (\alpha^{-1})^n$ is an inferior factor corresponding to, say, the positive solution a, b, and $\alpha^{-n} = a - \sqrt{D}b$. We will show that $(u + \sqrt{D}v)\alpha^{-n}$ is a dominant factor corresponding to some positive solution; hence (6) will contradict the minimality of α.

Let

$$c + \sqrt{D}d = (u + \sqrt{D}v)\alpha^{-n} = (u + \sqrt{D}v)(a - \sqrt{D}b).$$

It is easily verified by straightforward computation that $c = au - Dbv$, $d = av - bu$ is a solution of the Pell equation. It is a positive solution, because from (6), $c + \sqrt{D}d > 1$, and since c,d is a solution, the conjugate $c - \sqrt{D}d$ is positive and less than 1. Hence the sum of these factors is positive (which implies $c > 0$) and the difference $c + \sqrt{D}d - (c - \sqrt{D}d)$ is positive (which implies $d > 0$). Thus, we have our contradiction and the proof is complete.

EXERCISES

34-1. Suppose α, $\beta \in \mathcal{S}$, say $\alpha = x + \sqrt{D}y$ and $\beta = u + \sqrt{D}v$. If $x < u$, prove $y < v$. Hence, prove that $|\alpha - \beta| > 1$.

34-2. Find the fundamental solution of $x^2 - 11y^2 = 1$.

34-3. Prove that if $\alpha, \beta \in \mathcal{S}$, then $(\alpha^{-1}\beta^{-1})^{-1} \in \mathcal{S}$. Show that this proves the product of inferior factors corresponding to positive solutions is an inferior factor corresponding to a positive solution.

34-4. Suppose β is a dominant factor and γ is an inferior factor. Prove
   (a) $\beta\gamma$ is a dominant factor if and only if $\beta > \gamma^{-1}$;
   (b) $\beta\gamma$ is an inferior factor if and only if $\beta < \gamma^{-1}$;
   (c) $\beta\gamma$ does not correspond to a positive solution if and only if $\beta = \gamma^{-1}$.

34-5. Prove that the set

$$\mathcal{T} = \mathcal{S} \cup \{\beta^{-1} : \beta \in \mathcal{S}\} \cup \{1\}$$

is a group under multiplication. The proof may be facilitated by first noticing that

$$\mathcal{T} = \{\beta^n : \beta \in \mathcal{S}, n \in \mathcal{Z}\}.$$

34-6. Suppose x,y and u,v are pairs of positive integers. We write $(x,y)\mathcal{R}(u,v)$ if and only if equation (5) is satisfied. Prove $\mathcal{R}$ is an equivalence relation on the set of pairs x,y such that $x^2 - Dy^2 = k$.

34-7. Suppose D is a prime. If -1 is a quadratic non-residue (mod D), prove $x^2 - Dy^2 = -1$ has no solutions.

34-8. If $x^2 - Dy^2 = -1$ has a solution, prove it has a positive solution x_1, y_1 for which $x_1 + \sqrt{D}y_1$ is minimal. Then prove that odd powers of $x_1 + \sqrt{D}y_1$ correspond to positive solutions of the given equation, and even powers of $x_1 + \sqrt{D}y_1$ correspond to positive solutions of the Pell equation.

35. Rational Approximations of Reals

In this section we use some of the results of the previous sections of this chapter to find rational numbers which are in some sense "close" to a given real number.

The following theorem which we saw earlier (Lemma 26.2) is proved here using only our results on Farey sequences.

Theorem 35.1. *If ξ is irrational and $n \in \mathscr{Z}^+$, there exist integers a,b such that $(a,b) = 1$ and*

$$\left| \xi - \frac{a}{b} \right| < \frac{1}{b(n + 1)}, \qquad 0 < b \leq n.$$

Proof. We first assume $0 < \xi < 1$. In the Farey sequence $\mathscr{F}_n$, find the unique pair of successors $a/b < c/d$ such that ξ is between them. Since these are successive, evidently their mediant $(a + c)/(b + d)$ is not in $\mathscr{E}_n$, and $b + d \geq n + 1$. Let us assume ξ is between a/b and the mediant; the arguments are the same if ξ is between the mediant and c/d. Then

$$\left| \xi - \frac{a}{b} \right| < \left| \frac{a + c}{b + d} - \frac{a}{b} \right| = \frac{1}{b(b + d)} \leq \frac{1}{b(n + 1)},$$

and since $a/b \in \mathscr{E}_n$, $0 < b \leq n$, and $(a,b) = 1$. This proves the theorem if $0 < \xi < 1$.

If ξ is not between 0 and 1, write $\xi = [\xi] + \theta$, $0 < \theta < 1$. Let x,y be the relatively prime integers such that

$$\left| \theta - \frac{x}{y} \right| < \frac{1}{y(n + 1)}, \qquad 0 < y \leq n.$$

Then

$$\left| \xi - \frac{y[\xi] + x}{y} \right| = \left| \xi - [\xi] - \frac{x}{y} \right| = \left| \theta - \frac{x}{y} \right| < \frac{1}{y(n + 1)},$$

and we take $a = y[\xi] + x$, $b = y$. Clearly $0 < b \leq n$, and $(a,b) = (x,y) = 1$.

Theorem 35.2. *If ξ is irrational, there are infinitely many $a,b \in \mathscr{Z}$ such that $(a,b) = 1$ and*

$$\left| \xi - \frac{a}{b} \right| < \frac{1}{b^2}.$$

This was proved earlier as a corollary (32.1) to results on the convergents to an irrational ξ. Also, notice that this theorem is obtained as a weak corollary to Theorem 35.1 in this way: For $n_1 = 1$, find a',b' such that

$$\left| \xi - \frac{a'}{b'} \right| < \frac{1}{b'(n_1 + 1)} < \frac{1}{(b')^2}.$$

Now choose $n_2 \in \mathscr{Z}^+$ such that

$$\frac{1}{n_2} < \left| \xi - \frac{a'}{b'} \right|;$$

then there exist a'',b'' such that

$$\left| \xi - \frac{a''}{b''} \right| < \frac{1}{b''(n_2 + 1)} < \frac{1}{(b'')^2},$$

and since

$$\left| \xi - \frac{a''}{b''} \right| < \frac{1}{b''(n_2 + 1)} < \frac{1}{n_2} < \left| \xi - \frac{a'}{b'} \right|,$$

the pair a'',b'' is different from a',b'. Choose n_3 such that

$$\frac{1}{n_3} < \left| \xi - \frac{a''}{b''} \right|$$

and continue by induction.

Theorem 35.3. (*Hurwitz.*) *If ξ is irrational, there exist infinitely many $x,y \in \mathscr{Z}$ such that*

$$\left| \xi - \frac{x}{y} \right| < \frac{1}{\sqrt{5}y^2}.$$

Proof. We again prove the theorem in case $0 < \xi < 1$; the extension is exactly the same as in the proof of Theorem 35.1.

Let $n \in \mathscr{Z}^+$ and find the successive fractions $a/b < c/d$ in $\mathscr{F}_n$ which surround ξ. If we assume that neither a/b nor c/d satisfies the inequality in the theorem, then we have by assumption

(7) $$\xi - \frac{a}{b} \geq \frac{1}{\sqrt{5}b^2} \quad \text{and} \quad \frac{c}{d} - \xi \geq \frac{1}{\sqrt{5}d^2}.$$

Adding these inequalities, we obtain

$$\frac{c}{d} - \frac{a}{b} \geq \frac{1}{\sqrt{5}} \left(\frac{1}{b^2} + \frac{1}{d^2} \right);$$

by using Corollary 33.1a this can be put in the form

(8) $$b^2 - \sqrt{5}bd + d^2 \leq 0.$$

Now we use the mediant $(a + c)/(b + d)$; either ξ is between a/b and the mediant, or ξ is between the mediant and c/d. Let us suppose first that

$$\frac{a}{b} < \xi < \frac{a + c}{b + d}$$

and suppose that the mediant also fails to satisfy the theorem, so that

$$\frac{a + c}{b + d} - \xi \geq \frac{1}{\sqrt{5}(b + d)^2}.$$

Adding corresponding members of this inequality to the first inequality in (7), we can put the result in the form

(9) $$b^2 - \sqrt{5}b(b + d) + (b + d)^2 \leq 0.$$

We will show that (8) and (9) cannot be satisfied simultaneously by $b, d \in \mathscr{Z}^+$.

Divide both sides of (8) by b^2 and then let $u = d/b$ to get

(10) $$u^2 - \sqrt{5}u + 1 \leq 0, \qquad u = \frac{d}{b}.$$

Divide (9) by b^2 and let $u = 1 + d/b$ to get

(11) $$u^2 - \sqrt{5}u + 1 \leq 0, \qquad u = 1 + \frac{d}{b}.$$

Since $f(u) = u^2 - \sqrt{5}u + 1$ is a parabola which opens up, it is clear that $f(u) \leq 0$ if and only if u is between the u-intercepts of the graph of f, hence

$$f(u) \leq 0 \quad \text{if and only if} \quad \frac{\sqrt{5} - 1}{2} \leq u \leq \frac{\sqrt{5} + 1}{2}.$$

Consequently, (10) and (11) are satisfied if and only if

$$\frac{\sqrt{5} - 1}{2} \leq \frac{d}{b} \leq \frac{\sqrt{5} + 1}{2} \quad \text{and} \quad \frac{\sqrt{5} - 1}{2} \leq 1 + \frac{d}{b} \leq \frac{\sqrt{5} + 1}{2}.$$

These conditions are both satisfied only for $d/b = (\sqrt{5} - 1)/2$, which is impossible because d/b is rational and $(\sqrt{5} - 1)/2$ is irrational. Therefore, in the case when ξ is between a/b and the mediant, at least one of our assumptions is untenable, so at least one of $a/b, (a + c)/(b + d), b/d$ satisfies the inequality of the theorem.

In case ξ is between the mediant and c/d, we again assume the mediant fails to satisfy the theorem and we have

$$\xi - \frac{a + c}{b + d} \geq \frac{1}{\sqrt{5}(b + d)^2}.$$

Adding this to the second inequality in (7), we put the result in the form

(12) $$d^2 - \sqrt{5}d(b + d) + (b + d)^2 \leq 0.$$

Upon dividing through by d^2 in (8) and (12), we get $u^2 - \sqrt{5}u + 1 \leq 0$, with $u = b/d$ and $u = 1 + b/d$, respectively. Arguing as in the previous case, we find that $b/d = (\sqrt{5} - 1)/2$, again a contradiction.

We have now proved that if $a/b < c/d$ are successive in $\mathscr{F}_n$ and if $a/b < \xi < c/d$, then one of the numbers $a/b, (a + c)/(b + d), c/d$ gives a solution x/y of the inequality stated in the theorem, and we can inductively describe a construction which provides infinitely many solutions. Suppose for definiteness that

$$\frac{a + c}{b + d} < \xi < \frac{c}{d}.$$

In Exercise 33-4, we take $x' = (a + c)/(b + d)$ and $x'' = \xi$ to find a Farey sequence $\mathscr{F}_{m'}$ containing a'/b' between x' and x''; using Exercise 33-4 again with $x' = \xi$ and $x'' = c/d$ we find $\mathscr{F}_{m''}$ containing c'/d' between x' and x''. If $m = \max(m', m'')$, then $\mathscr{F}_m$ contains both a'/b' and c'/d'; by relabeling the fractions, if necessary, we can take a'/b' to be the largest fraction in $\mathscr{F}_m$ which does not exceed ξ, and c'/d' to be the smallest in $\mathscr{F}_m$ which ξ does not exceed. Hence $a'/b' < c'/d'$ are successive and ξ is between them; therefore one of $a'/b', (a' + c')/(b' + d'), c'/d'$ gives a solution of the inequality, and this new solution is necessarily different from the previous solution.

We will now prove that all rational numbers which are close (in the sense of Theorem 35.4) to an irrational ξ are convergents to ξ. We require the following.

Lemma 35.4. *Suppose ξ is irrational and P_k/Q_k is the kth convergent to ξ. If $a \in \mathscr{Z}, b \in \mathscr{Z}^+$, and for some $n \geq 0, b < Q_{n+1}$, then $|b\xi - a| \geq |Q_n b - P_n|$.*

Proof. Consider the system of equations

(13) $$\begin{cases} xP_n + yP_{n+1} = a \\ xQ_n + yQ_{n+1} = b. \end{cases}$$

The unique solution is $x = \Delta(n)(aQ_{n+1} - bP_{n+1}) = \pm(aQ_{n+1} - bP_{n+1})$, $y = \pm(bP_n - aQ_n)$; here we have used Theorem 31.3. Notice that $x, y \in \mathscr{Z}$.

Clearly $x \neq 0$, because if $x = 0$, then $aQ_{n+1} = bP_{n+1}$, and $(Q_{n+1}, P_{n+1}) = 1$ implies $Q_{n+1}|b$, so $Q_{n+1} \leq b$, a contradiction. Hence $|x| \geq 1$; if $y = 0$, the proof is complete because $y = 0$ implies $xQ_n = b$ and $xP_n = a$ by (13); therefore, $|b\xi - a| = |x||Q_n\xi - P_n| \geq |Q_n\xi - P_n|$. If $y \neq 0$, then $xy < 0$, because if $x < 0$, then $0 > xQ_n = b - yQ_{n+1}$, so $y > b/Q_{n+1} > 0$; and if $x > 0$, then $yQ_{n+1} = b - xQ_{n+1} < b < Q_{n+1}$, so $y < 0$.

Thus, if $y \neq 0$, x and y have opposite signs. Also $Q_{n+1}\xi - P_{n+1}$ and $Q_n\xi - P_n$ have opposite signs because ξ is between the two consecutive convergents P_{n+1}/Q_{n+1} and P_n/Q_n. Therefore the numbers $x(Q_n\xi - P_n)$ and $y(Q_{n+1}\xi - P_{n+1})$ have the same sign. Consequently, the absolute value of their sum is the sum of their absolute values, and

$$|b\xi - a| = |(xQ_n + yQ_{n+1})\xi - (xP_{n+1} + yP_{n+1})|$$
$$= |x(Q_n\xi - P_n) + y(Q_{n+1}\xi - P_{n+1})|$$
$$= |x||Q_n\xi - P_n| + |y||Q_{n+1}\xi - P_{n+1}|$$
$$> |Q_n\xi - P_n|.$$

Theorem 35.4. *If ξ is irrational and if there exist $a,b \in \mathscr{Z}$, $b \geq 1$, such that*

$$\left| \xi - \frac{a}{b} \right| < \frac{1}{2b^2},$$

then a/b is a convergent to ξ.

Proof. We may assume $(a,b) = 1$ and suppose a/b is not a convergent to ξ. Then we can find the unique n such that $Q_n \leq b < Q_{n+1}$. For this n, $|Q_n\xi - P_n| \leq |b\xi - a| < 1/2b$, so

$$\left| \xi - \frac{P_n}{Q_n} \right| < \frac{1}{2bQ_n}.$$

Since $a/b \neq P_n/Q_n$, we have $bP_n - aQ_n$ is a non-zero integer, and

$$\frac{1}{bQ_n} \leq \frac{|bP_n - aQ_n|}{bQ_n} = \left| \frac{P_n}{Q_n} - \frac{a}{b} \right| \leq \left| \frac{P_n}{Q_n} - \xi \right| + \left| \xi - \frac{a}{b} \right| < \frac{1}{2bQ_n} + \frac{1}{2b^2};$$

this leads to the contradiction $b < Q_n$, and the proof is complete.

Corollary 35.4a. *Every solution x/y of the inequality in the Hurwitz Theorem is a convergent to ξ.*

Corollary 35.4b. *If x,y is a positive solution of $x^2 - Dy^2 = 1$, then x/y is a convergent to $\sqrt{D}$.*

Proof. Since

$$|x - \sqrt{D}y| = \frac{1}{|x + \sqrt{D}y|} < \frac{1}{x + y},$$

we have

$$\left| \sqrt{D} - \frac{x}{y} \right| < \frac{1}{y(x+y)} < \frac{1}{2y^2},$$

since

$$x - y > x - \sqrt{D}y > 0.$$

We now prove that the constant $\sqrt{5}$ in the Hurwitz Theorem is best possible.

Theorem 35.5. *If $c > \sqrt{5}$, then there exists an irrational number x such that*

$$\left| \xi - \frac{a}{b} \right| < \frac{1}{cb^2}$$

is satisfied by only a finite number of a/b.

Proof. If $c > \sqrt{5} > 2$, then $|\xi - a/b| < 1/cb^2 < 1/2b^2$ can be satisfied only if a/b is a convergent to ξ. Take $\xi = (\sqrt{5}+1)/2 = \| 1,1,1,\dots \|$. By Exercise 32-6, $\lim_{n \to \infty} (\theta^{-1} + Q_{n-1}/Q_n) = \sqrt{5}$, so for only finitely many n can we have $\theta^{-1} + Q_{n-1}/Q_n > c$. Hence

$$\left| \xi - \frac{P_n}{Q_n} \right| = \left| \frac{\theta_n^{-1} P_n + P_{n-1}}{\theta_n^{-1} Q_n + Q_{n-1}} - \frac{P_n}{Q_n} \right| = \frac{1}{Q_n^2 \left(\theta_n^{-1} + \dfrac{Q_{n-1}}{Q_n} \right)} < \frac{1}{cQ_n^2}$$

can hold for only a finite number of convergents.

EXERCISES

35-1. In $\mathscr{F}_1$, $(\sqrt{5}-1)/2$ is between $1/2$ and $1/1$. Find a solution of

$$\left| \frac{\sqrt{5}-1}{2} - \frac{a}{b} \right| < \frac{1}{\sqrt{5}b^2}.$$

Use this to find a solution of

$$\left| \frac{\sqrt{5}+5}{2} - \frac{a}{b} \right| < \frac{1}{\sqrt{5}b^2}.$$

35-2. In $\mathscr{F}_6$, $2/5 < 1/2$ are successive and surround $1/\sqrt{6}$. Find a solution of $|1/\sqrt{6} - a/b| < 1/\sqrt{5}b^2$ among the candidates $2/5$, $3/7$, $1/2$. Now follow the construction defined in the proof of the Hurwitz Theorem to find a solution different from the first.

35-3. Show that $|1/\sqrt{5} - 17/38| < 1/2(38)^2$. Now verify that $17/38$ is a convergent to $1/\sqrt{5}$.

Chapter 7

THE EQUATION $x^n + y^n = z^n$, $n \leq 4$

ONE OF THE most widely known conjectures for which no proof has yet been given is "Fermat's Last Theorem": The Diophantine equation $x^n + y^n = z^n$ has no solution with $xyz \neq 0$ if $n > 2$. In this chapter we will show that for $n = 2$ the equation has solutions and we will find all of them. Then we will show that for $n = 3$ and $n = 4$ there are no solutions.

36. Pythagorean Triples

If the integers x,y,z satisfy the equation

$$(1) \qquad x^2 + y^2 = z^2,$$

then x,y,z is said to be a *Pythagorean triple*. Suppose x,y,z is such a triple and $(x,y,z) = d$. If we put $x = dx_1$, $y = dy_1$, $z = dz_1$, we see that x_1,y_1,z_1 is also a Pythagorean triple and $(x_1,y_1,z_1) = 1$. On the other hand if x,y,z is any solution of (1) and $k \in \mathscr{Z}$, then kx,ky,kz is also a solution. Thus, any solution of (1) may be used to find a solution x,y,z such that $(x,y,z) = 1$, and conversely, a solution with $(x,y,z) = 1$ may be used to generate a family of solutions.

A solution x,y,z such that $(x,y,z) = 1$ is called a *primitive* Pythagorean triple. The above remarks indicate that it suffices to find all primitive solutions of (1). We confine our attention to the case $x > 0$, $y > 0$, $z > 0$.

If x,y,z is a primitive positive solution of (1), then $(x,y) = (x,z) = (y,z) = 1$. To see this, suppose $(x,y) = d$. Then $d^2|(x^2 + y^2) = z^2$, so $d|z$, and $d|(x,y,z)$. The arguments are similar to show $(x,z) = (y,z) = 1$. In particular, then, not both x and y can be even since $(x,y) = 1$. Also, not both can be odd because, if so, x^2 and y^2 are 1 (mod 4), and $2 \equiv x^2 + y^2 \equiv z^2$ (mod 4) is impossible. Therefore, x and y must have opposite parity; we assume x is even, y is odd.

Since x is even, write $x = 2x_1$. Then (1) may be written as

$$(2) \qquad z^2 - y^2 = (z - y)(z + y) = 4x_1^2.$$

Now since both y, z are odd, $z \pm y$ are even, say $z - y = 2r_1$ and $z + y = 2s_1$. Then (2) becomes

$$(3) \qquad r_1 s_1 = x_1^2.$$

Notice that

$$(r_1, s_1) = \left(\frac{z - y}{2}, \frac{z + y}{2} \right) \Bigg| \frac{(z - y) + (z + y)}{2} = z$$

and

$$(r_1, s_1) \Bigg| \frac{(z + y) - (z - y)}{2} = y.$$

Therefore $(r_1, s_1) = 1$, and because of (3), both r_1 and s_1 must be squares, say $r_1 = r^2$ and $s_1 = s^2$. It is easy to see that $z = s^2 + r^2$ and $y = s^2 - r^2$; since y, z are odd we must have s^2, r^2 (and hence s, r) of opposite parity.

Thus, every positive primitive solution of (1) implies the existence of positive integers $r < s$ which are relatively prime, of opposite parity, and such that $x = 2sr$, $y = s^2 - r^2$, and $z = s^2 + r^2$. Conversely, if r, s are as described here, then x, y, z is obviously a positive Pythagorean triple. Moreover, if $d = (x, y, z)$, let $e = (x, y)$. Clearly $d | e$. If $e > 1$, then e has a prime divisor p. Hence $p | (x, y) = (2sr, s^2 - r^2)$, and since $s^2 - r^2$ is odd, we know $p | (sr, s^2 - r^2)$ and $p \neq 2$. Then

$$(4) \qquad \begin{aligned} sr &\equiv 0 \, (\text{mod } p) \\ s^2 - r^2 &\equiv 0 \, (\text{mod } p). \end{aligned}$$

Since $(s, r) = 1$, we have exactly one of r, s divisible by p, and this is incompatible with (4). Therefore, $d = e = 1$ and x, y, z is primitive.

We have proved Theorem 36.1.

Theorem 36.1. *The positive primitive solutions of $x^2 + y^2 = z^2$ are $x = 2sr$, $y = s^2 - r^2$, $z = s^2 + r^2$, where $0 < r < s$, $(r, s) = 1$, and r, s are of opposite parity.*

EXERCISES

36-1. Determine all positive primitive Pythagorean triples x, y, z such that $x < y < z \leq 50$.

36-2. How many imprimitive positive Pythagorean triples are there which satisfy $x < y < z \leq 50$?

36-3. Find all right triangles with sides of integral lengths and with area 60.

36-4. If x,y,z is a primitive Pythagorean triple, prove that $x \pm y$ are either $\equiv 1 \pmod 8$ or $\equiv -1 \pmod 8$.

36-5. Suppose x,y,z is a Pythagorean triple, not necessarily primitive. Prove that at least one of x,y,z is divisible by 3, at least one by 4, and at least one by 5. Thus, $60|xyz$.

37. The Equation $x^4 + y^4 = z^4$

We will first prove a more general result than Fermat's Last Theorem for $n = 4$.

Theorem 37.1. *The equation $x^4 + y^4 = z^2$ has no solution in positive integers.*

Proof. Suppose there is a positive triple x,y,z such that $x^4 + y^4 = z^2$. We may assume that $(x,y) = 1$, and hence that $(x^2,y^2,z) = 1$. Further, we may assume that z is the least positive integer for which there exists $x > 0$, $y > 0$ such that

$$(5) \qquad x^4 + y^4 = z^2, \qquad (x,y) = 1.$$

Under these conditions, by Theorem 36.1 there are positive integers $r < s$, $(r,s) = 1$, with r and s of opposite parity, and

$$x^2 = 2sr$$
$$y^2 = s^2 - r^2$$
$$z = s^2 + r^2.$$

If s is even, r odd, then $y^2 = -1 \pmod 4$, which is impossible. Thus, s is odd and r is even, say $r = 2r'$, and we have

$$x^2 = 4sr'.$$

Since $(r',s) = 1$, both r' and s are squares, say $r' = r_1^2$ and $s = s_1^2$. Then from $y^2 + r^2 = s^2$ we get

$$y^2 + (2r_1^2)^2 = (s_1^2)^2.$$

Again by Theorem 36.1, there are integers $0 < r_2 < s_2, (r_2,s_2) = 1$, such that

$$2r_1^2 = 2r_2 s_2$$
$$y = s_2^2 - r_2^2$$
$$s_1^2 = s_2^2 + r_2^2.$$

Since $r_1^2 = r_2 s_2$ and $(r_2, s_2) = 1$, both r_2 and s_2 are squares, say $r_2 = r_3^2$ and $s_2 = s_3^2$. Hence

(6) $$s_1^2 = s_3^4 + r_3^4, \qquad (s_3, r_3) = 1.$$

But $0 < s_1 < s_1^2 = s < s^2 + r^2 = z$, so (6) contradicts the way z was chosen in (5).

Corollary 37.1. *The equation* $x^4 + y^4 = z^4$ *has no solution with* $xyz \ne 0$.

Proof. If this equation had a solution x, y, z such that $xyz \ne 0$, then $|x|, |y|, z^2$ would be a positive solution of the equation in Theorem 37.1.

EXERCISES

37-1. Suppose $m \in \mathscr{Z}^+$. Prove the equation $x^{4m} + y^{4m} = z^{4m}$ has no solution in positive integers.

37-2. Show that in order to prove Fermat's Last Theorem, it suffices to prove that $x^p + y^p = z^p$ has no solution with $xyz \ne 0$ for odd primes p.

38. Arithmetic in $\mathscr{K}(\sqrt{-3})$

Before we can prove Fermat's Last Theorem for $n = 3$, we must investigate some properties of the set

$$\mathscr{J} = \left\{ a + b\xi : a, b \in \mathscr{Z}, \xi = \frac{-1 + \sqrt{-3}}{2} \right\}$$

of "Jacobian integers" in the imaginary quadratic number field

$$\mathscr{K}(\sqrt{-3}) = \left\{ \frac{a}{c} + \frac{b}{d}\sqrt{-3} : a, b, c, d \in \mathscr{Z}, cd \ne 0 \right\}.$$

The discussion will be very similar to the discussion of Gaussian integers. Since $\mathscr{J}$ is contained in the set of complex numbers, two integers in $\mathscr{J}$ are added and multiplied as complex numbers; thus, if $\alpha, \beta \in \mathscr{J}$, then $\alpha \pm \beta$, $\alpha\beta \in \mathscr{J}$.

Notice that $\xi^2 = -1 - \xi$ and $\xi^3 = 1$. If $\alpha \in \mathscr{J}$, then the *norm* $D(\alpha)$ of α is defined to be $\alpha\bar{\alpha}$, where $\bar{\alpha}$ is the complex conjugate of α. Therefore, if $\alpha = a + b\xi$, then

$$D(\alpha) = D(a + b\xi) = \left(a - \frac{b}{2} + \frac{b\sqrt{-3}}{2} \right)\left(a - \frac{b}{2} - \frac{b\sqrt{-3}}{2} \right)$$

$$= a^2 - ab + b^2.$$

We have $D(\alpha\beta) = D(\alpha) D(\beta)$. An integer $\alpha \in \mathscr{J}$ is a *unit* if and only if $\alpha | 1$, hence if and only if $D(\alpha) = 1$. The units in $\mathscr{J}$ are ± 1, $\pm \xi$, $\pm \xi^2$, and no others. Two integers α, β are called *associates* if there is a unit γ such that $\alpha = \beta\gamma$. A non-zero, non-unit δ in $\mathcal{J}$ is called a *prime* if every factorization of δ involves an associate of δ. To describe the primes in $\mathcal{J}$ we need this lemma which gives some information about rational integers n which can be written in the form $n = x^2 + 3y^2$, $x, y \in \mathcal{Z}$.

Lemma 38.1. *If $n \in \mathcal{Z}^+$ and if there is a prime $q \in \mathcal{Z}$ such that $q|n$ and $q \equiv 5 \pmod{6}$, then n cannot be written in the form $n = x^2 + 3y^2$, $(x,y) = 1$, $x, y \in \mathcal{Z}$. Every prime $p \in \mathcal{Z}$, $p \equiv 1 \pmod{6}$ has a representation as $p = x^2 + 3y^2$ with $(x,y) = 1$, $x, y \in \mathcal{Z}$.*

Proof. Suppose there is a prime q such that $q|n$, $q \equiv 5 \pmod{6}$, and $n = a^2 + 3b^2$, $(a,b) = 1$, for some $a, b \in \mathcal{Z}$. Obviously $q \nmid b$, so there exists x such that $bx \equiv 1 \pmod{q}$. Then

$$0 \equiv x^2 n = x^2(a^2 + 3b^2) \equiv (xa)^2 + 3 \pmod{q}.$$

But then $(-3/q) = 1$ and this contradicts the result in Exercise 17-4.

Now suppose $p \in \mathcal{Z}$ is prime and $p \equiv 1 \pmod{6}$. This proof closely parallels the proof that primes congruent to 1 (mod 4) can be written as a sum of squares, and we again use the approximation of Lemma 26.2. We know (from Exercise 17-4) that $(-3/p) = 1$, so there is some u such that $u^2 \equiv -3 \pmod{p}$. Let $z = -u/p$ and $n = [\sqrt{pk}]$, where $k = 1/\sqrt{3}$. There are $a, b \in \mathcal{Z}$ such that

$$\left| -\frac{u}{p} - \frac{a}{b} \right| \le \frac{1}{b(n+1)} < \frac{1}{b\sqrt{pk}}, \qquad 1 \le b \le \sqrt{pk}.$$

Take $c = ub + ap$, and find that

$$|c| = \left| -\frac{u}{p} - \frac{a}{b} \right| |pb| < \sqrt{\frac{p}{k}}.$$

Now $0 < c^2 + 3b^2 < p/k + 3pk = 2\sqrt{3}p < 4p$, and $c^2 + 3b^2 \equiv u^2b^2 + 3b^2 = b^2(u^2 + 3) \equiv 0 \pmod{p}$. Therefore, we have either $c^2 + 3b^2 = p, 2p$, or $3p$.

If $c^2 + 3b^2 = p$ we are through. It is easy to see that $c^2 + 3b^2 = 2p$ is impossible, because if b and c have the same parity then $4|c^2 + 3b^2$, but $4 \nmid 2p$; also, b and c cannot have opposite parity because then $c^2 + 3b^2$ is odd, but $2p$ is even. Finally, if $c^2 + 3b^2 = 3p$, then $3|c$, say $3c_1 = c$, and we find that $3c_1^2 + b^2 = p$. In any case we have $p = x^2 + 3y^2$. If $(x,y) = d$, then $d^2|p$, so $d = 1$.

Theorem 38.1. *The primes in $\mathcal{J}$ are 2, $1 - \xi$, the rational primes $q \equiv 5 \pmod{6}$, and the integers $a + b + 2b\xi$ and $a - b - 2b\xi$, where $a^2 + 3b^2 = p$ is a rational prime congruent to 1 (mod 6), and associates of these.*

Proof. We notice first that if $D(\alpha)$ is prime in $\mathscr{Z}$, then α is prime in $\mathscr{J}$. Also, just as in the case of Gaussian integers, corresponding to each prime α in $\mathscr{J}$, there is one and only one rational prime p such that $\alpha\beta = p$ for some $\beta \in \mathscr{J}$. (The proof of this fact in $\mathscr{J}$, as in $\mathscr{G}$, depends upon the important lemma that if α is prime and $\alpha|\beta\gamma$, then either $\alpha|\beta$ or $\alpha|\gamma$. We assume for now that this is true; an indication of the proof will be given shortly.) We have $p^2 = D(p) = D(\alpha)D(\beta)$. Thus, either $D(\alpha) = p$, or $D(\alpha) = p^2$ and α is an associate of p.

Let $\alpha = a + b\xi$. If $p = 2$, we see that $D(\alpha) = a^2 - ab + b^2 = 2$ is impossible, because a,b must both be even, but then $4|D(\alpha)$ and $4\nmid 2$. Therefore, the prime α is an associate of 2 in $\mathscr{J}$.

If $p = 3$, then α cannot be an associate of 3, since $3 = (1 - \xi)(2 + \xi)$ is not prime in $\mathscr{J}$. However, $D(1 - \xi) = D(2 + \xi) = 3$ implies that both $1 - \xi$ and $2 + \xi$ are prime in $\mathscr{J}$, and since $-\xi^2(1 - \xi) = 2 + \xi$, they are associates.

If $p \equiv 5 \pmod 6$, suppose $D(\alpha) = p$. Then $D(\alpha) = D(a + b\xi) = p = a^2 - ab + b^2$. If $(a,b) = d$, then $d^2|p$, so $(a,b) = 1$. Then also $(2a - b,b) = 1$, and $4p = (2a - b)^2 + 3b^2$ contradicts Lemma 38.1. Hence, $D(\alpha) \neq p$, so α is an associate of p.

If $p \equiv 1 \pmod 6$, we have seen that there are rational integers x,y such that $p = x^2 + 3y^2, (x,y) = 1$. Then

$$p = (x + y + 2y\xi)(x - y - 2y\xi).$$

Since $D(x + y + 2y\xi) = D(x - y - 2y\xi) = p$, α is an associate of one of these factors. Finally, the factors of p are not associates, as is easily seen by considering $\pm\xi^k(x + y + 2y\xi)$, $k = 0,1,2$.

The proof of unique factorization in $\mathscr{J}$ again follows from a division algorithm. In the proof we use the fact that if $\beta = b + d\xi$, then $\bar{\beta} = b + d\xi^2$, and this holds for all complex numbers, not only for integers in $\mathscr{J}$.

Theorem 38.2. *If* $\alpha,\beta \in \mathscr{J}, \beta \neq 0$, *then there exist* $\lambda,\theta \in \mathscr{J}$ *such that* $\alpha = \beta\lambda + \theta$, $D(\theta) < D(\beta)$.

Proof. If $\alpha,\beta \in \mathscr{J}, \beta \neq 0$, say $\alpha = a + c\xi$ and $\beta = b + d\xi$. Then

$$\frac{\alpha}{\beta} = \frac{(a + c\xi)(b + d\xi^2)}{(b + d\xi)(b + d\xi^2)} = \frac{(ab + cd - ad) + (cb - ad)\xi}{b^2 - bd + d^2}$$

$$= x + y\xi, \quad x,y \text{ rational.}$$

Choose $m,n \in Z$ such that $|x - m| \leq 1/2, |y - n| \leq 1/2$. Take $\lambda = m + n\xi$ and $\theta = \alpha - \beta\lambda$. Then

$$|\theta| = |\beta|\left|\frac{\alpha}{\beta} - \lambda\right| = |\beta||(x + y\xi) - (m + n\xi)| = |\beta||(x - m) + (y - n)\xi|$$

$$= |\beta|\{((x - m) + (y - n)\xi)((x - m) + (y - n)\xi^2)\}^{1/2}$$

$$= |\beta|\{(x - m)^2 + (y - n)^2 - (x - m)(y - n)\}^{1/2}$$

$$\leq |\beta|(\tfrac{1}{4} + \tfrac{1}{4} + \tfrac{1}{2} \cdot \tfrac{1}{2})^{1/2} < |\beta|.$$

Therefore, $D(\theta) < D(\beta)$.

With the division algorithm we can define a Euclidean algorithm, a gcd which is unique up to associates, and then prove that if a prime $\alpha \in \mathscr{J}$ divides $\beta\lambda$, then either $\alpha|\beta$ or $\alpha|\lambda$. Then the existence and essential uniqueness of factorization of non-zero, non-units into primes in $\mathscr{J}$ can be verified.

EXERCISES

38-1. Show that $\alpha,\beta \in \mathscr{J}$ implies $\alpha \pm \beta$, $\alpha\beta \in \mathscr{J}$.

38-2. Show that the units in $\mathscr{J}$ are $(-\xi)^k$, $k = 1, 2, \ldots, 6$.

38-3. Prove

    (a) $D(\bar{\alpha}) = D(\alpha)$.

    (b) $D(n\alpha) = n^2 D(\alpha)$ for all $n \in \mathscr{Z}^+$.

38-4. If $(\alpha,\beta) = \lambda$, show that λ is not unique but is an associate of any other gcd of α,β.

38-5. Factor $408 + 306\xi$ and $6 + 11\xi$ into products of primes in $\mathscr{J}$.

38-6. (See Exercise 28-3.) Prove that the rational prime p

    (a) ramifies in $\mathscr{J}$ if and only if $(-3/p)$ is undefined;

    (b) stays prime in $\mathscr{J}$ if and only if $(-3/p) = -1$;

    (c) splits in $\mathscr{J}$ if and only if $(-3/p) = 1$.

38-7. Prove that if α is prime in $\mathscr{J}$ and $\alpha|\beta\lambda$, then $\alpha|\beta$ or $\alpha|\lambda$.

39. The Equation $x^3 + y^3 = z^3$

To prove that the Fermat equation is impossible for $n = 3$, we will show that the equation $\alpha^3 + \beta^3 = \gamma^3$ is impossible for $\alpha,\beta,\gamma \in \mathscr{J}$ if $\alpha\beta\gamma \neq 0$. The prime $\pi = 1 - \xi$ plays an important role in what follows.

Lemma 39.1. *If $\alpha \in \mathscr{J}$ and $\pi\nmid\alpha$, then either $\pi^4|(\alpha^3 + 1)$ or $\pi^4|(\alpha^3 - 1)$.*

Proof. Recall that $\pi\bar{\pi} = 3 = -\xi^2\pi^2$. If $\alpha \in \mathscr{J}$, $\alpha = a + b\xi = a + b - b\pi$. Since $a + b \in \mathscr{Z}$, we find $q,r \in \mathscr{Z}$ such that $a + b = 3q + r$, $r = 0, \pm 1$. Thus,

$$\alpha = r + \pi(-\xi^2\pi q - b),$$

and in general we notice that *every $\beta \in \mathscr{J}$ can be written in the form $\beta = \pi\varepsilon + r$,* with $r = 0, 1,$ or -1. But since $\pi\nmid\alpha$, $r = 0$ is not possible; therefore, with an obvious definition for $\sigma \in \mathscr{J}$, we have $\alpha = \pm 1 + \pi\sigma$. Then

$$\alpha^3 = \pm 1 + 3\pi\sigma \pm 3\pi^2\sigma^2 + \pi^3\sigma^3$$

$$= \pm 1 \pm (-\xi^2)\pi^4\sigma^2 + \pi^3\sigma(\sigma^2 - \xi^2).$$

Evidently, to complete the proof we need only show that $\pi | \sigma(\sigma^2 - \xi^2)$. But this is obvious because if $\sigma = \pi\varepsilon + r$ with $r = 0$, then $\pi | \sigma$, and with $r = \pm 1$, $\pi | (\sigma \mp \xi)$. Therefore, π^4 divides $\alpha^3 + 1$ or $\alpha^3 - 1$.

Theorem 39.1. *If* $\alpha^3 + \beta^3 = \gamma^3$, $\alpha, \beta, \gamma \in \mathscr{J}$, *then* π *must divide at least one of* α, β, γ.

Proof. Suppose the contrary. Then by Lemma 39.1 we can write

$$\alpha^3 = \pm 1 + \pi^4 \sigma_1$$

$$\beta^3 = \pm 1 + \pi^4 \sigma_2$$

$$\gamma^3 = \pm 1 + \pi^4 \sigma_3.$$

Therefore, with $\sigma = \sigma_3 - \sigma_1 - \sigma_2$, we have

$$\pm 1 \pm 1 = \pm 1 + \pi^4 \sigma.$$

With the eight possible choices of sign considered, we get either $\pi^4 | 1$ or $\pi^4 | 3$. Neither is possible, since π cannot divide a unit and π^2 is an associate of 3.

Theorem 39.2. *The equation* $\gamma^3 = \alpha^3 + \beta^3$ *is impossible for* $\alpha, \beta, \gamma \in \mathscr{J}$ *if* $\alpha\beta\gamma \neq 0$.

Proof. Suppose there are non-zero integers $\alpha, \beta, \gamma \in \mathscr{J}$ such that $\alpha^3 + \beta^3 = \gamma^3$. If α, β, γ have a common factor, we can remove it to obtain a solution in which the integers have no common factor, and then, by familiar arguments, we can conclude that the new integers are relatively prime in pairs. Also, since π divides one of α, β, γ, we may suppose $\pi | \gamma$ [if not, say $\pi | \beta$, we can write $\beta^3 = \gamma^3 + (-\alpha)^3$], and furthermore we may suppose that the power of π in γ is the least positive power which divides any non-zero integer in $\mathscr{J}$ whose cube is a sum of two cubes. Thus we are assuming that

(7)
$$\gamma^3 = \alpha^3 + \beta^3, \quad (\alpha, \beta) = (\alpha, \gamma) = (\beta, \gamma) = 1, \quad \pi | \gamma,$$
and γ contains the minimal power of π.

We consider the factorization

$$\gamma^3 = (\alpha + \beta)(\alpha + \beta\xi)(\alpha + \beta\xi^2).$$

Since $\pi | \gamma$, π divides at least one of the factors $\alpha + \beta\xi^k$, $k = 0, 1, 2$. But $(\alpha + \beta) - (\alpha + \beta\xi) = \beta\pi$, and $(\alpha + \beta\xi) - (\alpha + \beta\xi^2) = \beta\xi\pi$, and since π divides one of the factors, it divides all three. Therefore, π divides all of

$$\alpha + \beta\xi, \, \alpha\xi + \beta = \xi(\alpha + \beta\xi^2), \, \alpha\xi^2 + \beta\xi^2$$

since these are associates of the original three factors. Also

$$(\alpha + \beta\xi) + (\alpha\xi + \beta) + (\alpha\xi^2 + \beta\xi^2) = (\alpha + \beta)(1 + \xi + \xi^2) = 0.$$

We now take

$$\alpha_1 = (\alpha + \beta\xi)/\pi, \qquad \beta_1 = (\alpha\xi + \beta)/\pi, \qquad \gamma_1 = (\alpha\xi^2 + \beta\xi^2)/\pi.$$

Suppose $(\alpha_1,\beta_1) = \delta$. Then $\delta|(\xi^2\beta_1 - \xi\alpha_1) = \alpha$ and $\delta|(\xi^2\alpha_1 - \xi\beta_1) = \beta$, so $\delta = 1$. Since $\alpha_1 + \beta_1 + \gamma_1 = 0$, we also have $(\alpha_1,\gamma_1) = 1$ and $(\beta_1,\gamma_1) = 1$.

From the relation $\alpha_1\beta_1\gamma_1 = (\gamma/\pi)^3$ and the pairwise relative primality of $\alpha_1,\beta_1,\gamma_1$, we can conclude that each is a cube, say

$$\alpha_1 = \alpha_2^3, \qquad \beta_1 = \beta_2^3, \qquad \gamma_1 = \gamma_2^3.$$

Evidently $(-\gamma_2)^3 = \alpha_2^3 + \beta_2^3$ and one of $\alpha_2,\beta_2,\gamma_2$ is divisible by π. Say $\pi|\gamma_2$. (It is in fact γ_2 which is a multiple of π, but we need not show this.) It remains only to show that the power of π in γ_2 is less than the power in γ. But this is obvious because $\gamma_2|\gamma_1|(\gamma/\pi)$, and we have constructed a triple which contradicts (7).

Corollary 39.2. *The equation* $x^3 + y^3 = z^3$ *has no solution in rational integers* x,y,z *such that* $xyz \neq 0$.

Chapter 8

THE PRIME NUMBER THEOREM

40. Introductory Remarks

If the function $\pi(x)$, x real, is defined by

$$\pi(x) = \mathbf{v}\{p : p \text{ is prime, } p \leq x\},$$

then from Exercise 4-7 we see that $\lim_{x \to \infty} \pi(x) = \infty$. In this chapter we will study the function $\pi(x)$ in some greater detail and we will prove the Prime Number Theorem:

$$\pi(x) \sim \frac{x}{\log x}.$$

This theorem was first proved in 1896 by Hadamard and Vallée de la Poisson independently; both proofs involve the theory of functions of a complex variable. In 1948, Atle Selberg and Paul Erdös, working independently, gave proofs of the Prime Number Theorem which were "elementary" in the sense that the proofs did not use the theory of functions.

In Selberg's first proof the concepts of limit inferior and limit superior played a basic role. In a second proof, which appeared in 1949, Selberg used nothing deeper than some simple properties of the logarithm function. Both proofs involve the function

$$\theta(x) = \sum_{p \leq x} \log p,$$

the summation extending over all *primes* $p \leq x$; for example, $\theta(9) = \log 2 + \log 3 + \log 5 + \log 7$. A modification of Selberg's proof was presented in 1951 by Harold Shapiro, and involves instead the function

$$\psi(x) = \sum_{p^\beta \leq x} \log p,$$

129

the summation including one term $\log p$ for each *power* of p which is $\leq x$; for example, $\psi(9) = 3 \log 2 + 2 \log 3 + \log 5 + \log 7$.

It is this latter proof which we will present in this chapter, because this proof depends only upon techniques already introduced and used in Chapter 2 and Chapter 4. We will first prove that $\psi(x) \sim x$, and later we will show this is equivalent to the Prime Number Theorem.

41. Preliminary Results

The notation $f \cdot g$ to denote the Dirichlet convolution product of two arithmetic functions has already been introduced. We now define the natural product fg and sum $f + g$ of two functions of an integral or real variable by

$$fg(x) = f(x)\, g(x)$$

$$(f + g)(x) = f(x) + g(x).$$

For example, with these notations we would have $\iota_0 \cdot \mu = \varepsilon$, $\iota_0 \mu = \mu$, and $(\iota_0 + \mu)(n) = \mu(n) + 1$ for every $n \in \mathscr{Z}^+$.

Throughout this chapter, we will write

$$\log x = L(x), \qquad x > 0.$$

We will use both the notations "log" and "L" for this function, depending upon which notation seems more convenient for a particular discussion.

The function $\Lambda(n) \in \mathscr{A}$ is defined at each $n \in \mathscr{Z}^+$ by

$$\Lambda(n) = \begin{cases} \log p & \text{if } n = p^\beta, \beta \geq 1, \\ 0 & \text{otherwise.} \end{cases}$$

Here, of course, p denotes a prime. It is easy to see that $\psi(x)$ is the summatory function of $\Lambda(n)$, that is,

$$\psi(n) = \sum_{n \leq x} \Lambda(n).$$

Lemma 41.1. *The following functional relations are valid*:
(1) $L = \Lambda \cdot \iota_0$;
(2) $\mu \cdot L = \Lambda$;
(3) $(-\mu L) \cdot \iota_0 = \Lambda$;
(4) $\mu \cdot (LL) = (\Lambda L) + (\Lambda \cdot \Lambda)$.

Proof. If

$$n = \prod_{i=1}^{k} p_i^{\beta_i},$$

then

$$L(n) = L\left(\prod_{i=1}^{k} p_i^{\beta_i}\right) = \sum_{i=1}^{k} L(p_i^{\beta_i}) = \sum_{i=1}^{k} \beta_i L(p_i)$$

$$= \sum_{i=1}^{k} \sum_{\theta_i=0}^{\beta_i} \Lambda(p_i^{\theta_i}) = \sum_{d|n} \Lambda(d) = \Lambda \cdot \iota_0(n).$$

This proves (1), and (2) follows by Möbius inversion.

Then

$$\Lambda(n) = \mu \cdot L(n) = \sum_{d|n} \mu(d) L\left(\frac{n}{d}\right)$$

$$= \sum_{d|n} \mu(d)\{L(n) - L(d)\}$$

$$= L(n) \sum_{d|n} \mu(d) - \sum_{d|n} \mu(d) L(d)$$

$$= L(n) \, \varepsilon(n) - (\mu L) \cdot \iota_0(n);$$

but $L(n) = 0$ if $n = 1$ and $\varepsilon(n) = 0$ if $n > 1$. Therefore, for all $n \in \mathscr{Z}^+$, $\Lambda(n) = (-\mu L) \cdot \iota_0(n)$, which proves (3).

Finally,

$$\mu \cdot (LL)(n) = \sum_{d|n} \mu(d) L\left(\frac{n}{d}\right) L\left(\frac{n}{d}\right)$$

$$= \sum_{d|n} \mu(d) L\left(\frac{n}{d}\right)\{L(n) - L(d)\}$$

$$= L(n)(\mu \cdot L)(n) - (\mu L) \cdot L(n)$$

$$= L(n) \Lambda(n) - (\mu L) \cdot \iota_0 \cdot \Lambda(n)$$

$$= (L\Lambda)(n) + (\Lambda \cdot \Lambda)(n)$$

and this proves (4).

We now would like an asymptotic formula for the summatory function of the right member of (4). In order to get this, we require some lemmas which come as easy consequences of the Euler–McLaurin sum formula (Theorem 20.1) and the inversion technique of Theorem 23.1.

Lemma 41.2.

$$\sum_{n \le x} LL(n) = \sum_{n \le x} \log^2 n = x \log^2 x - 2x \log x + 2x + O(\log^2 x)$$

Proof. From Theorem 20.1,

$$\sum_{n \le x} \log^2 n = \int_1^x \log^2 t \, dt + 2 \int_1^x (t - [t]) \frac{\log t}{t} \, dt + 0(\log^2 x)$$

$$= (t \log^2 t - 2t \log t + 2t) \bigg]_1^x + 0\left(\int_1^x \frac{\log t}{t} \, dt\right) + 0(\log^2 x)$$

$$= x \log^2 x - 2x \log x + 2x + 0(1) + 0(\log^2 t]_1^x) + 0(\log^2 x).$$

Lemma 41.3.

$$\sum_{n \le x} L(n) = \sum_{n \le x} \log n = x \log x - x + 0(\log x)$$

The proof is left as an exercise.

Lemma 41.4.

$$\sum_{n \le x} \frac{\log n}{n} = \frac{1}{2} \log^2 x + C_3 + 0\left(\frac{\log x}{x}\right)$$

for some constant C_3.
   Proof. Again by Theorem 20.1,

$$\sum_{n \le x} \frac{\log n}{n} = \int_1^x \frac{\log t}{t} \, dt + \int_1^x \frac{(t - [t])(1 - \log t)}{t^2} \, dt + 0\left(\frac{\log x}{x}\right)$$

$$= \frac{1}{2} \log^2 t \bigg]_1^x + \int_1^\infty \frac{(t - [t])(1 - \log t)}{t^2} \, dt$$

$$- \int_x^\infty \frac{(t - [t])(1 - \log t)}{t^2} \, dt + 0\left(\frac{\log x}{x}\right)$$

if the improper integrals converge. But

$$\int_x^\infty \frac{(t - [t])(1 - \log t)}{t^2} \, dt = 0\left(\int_x^\infty \frac{1 - \log t}{t^2} \, dt\right) = 0\left(\frac{\log t}{t}\bigg]_x^\infty\right) = 0\left(\frac{\log x}{x}\right),$$

which approaches 0 as $x \to \infty$. Therefore, the integral from 1 to ∞ converges, say to C_3, and the proof is complete.

Lemma 41.5.

$$\sum_{n \le x} \frac{\mu(n)}{n} \log \frac{x}{n} = 0(1)$$

Proof. Take $g(x) = x$ in Theorem 23.1. Use Lemma 19.1 and Corollary 23.1. The details are left as an exercise.

Lemma 41.6.

$$\sum_{n \le x} \frac{\mu(n)}{n} \log^2 \frac{x}{n} = 2 \log x + O(1)$$

Proof. With $g(x) = x \log x$ in Theorem 23.1 we have

$$f(x) = \sum_{n \le x} g\left(\frac{x}{n}\right) = x \sum_{n \le x} \frac{1}{n}(\log x - \log n)$$

$$= x \log x \sum_{n \le x} \frac{1}{n} - x \sum_{n \le x} \frac{\log n}{n}$$

$$= x \log x \{\log x + O(1)\} - x\{\tfrac{1}{2} \log^2 x + O(1)\}$$

$$= \frac{x}{2} \log^2 x + O(x \log x).$$

Therefore,

$$x \log x = g(x) = \sum_{n \le x} \mu(n) f\left(\frac{x}{n}\right)$$

$$= \sum_{n \le x} \mu(n)\left\{\frac{x}{2n} \log^2 \frac{x}{n} + O\left(\frac{x}{n} \log \frac{x}{n}\right)\right\}$$

$$= \frac{x}{2} \sum_{n \le x} \frac{\mu(n)}{n} \log^2 \frac{x}{n} + O\left(x \sum_{n \le x} \frac{\mu(n)}{n} \log \frac{x}{n}\right)$$

$$= \frac{x}{2} S_1 + O(xO(1)) = \frac{x}{2} S_1 + O(x).$$

Hence, $S_1 = 2 \log x + O(1)$.

Theorem 41.1.

$$\sum_{n \le x} ((\Lambda L) + (\Lambda \cdot \Lambda))(n) = 2x \log x + O(x)$$

Proof. By Lemma 41.1(4) and Theorem 22.1,

$$\sum_{n \le x} ((\Lambda L) + (\Lambda \cdot \Lambda))(n) = \sum_{n \le x} \mu \cdot (LL)(n) = \sum_{n \le x} \mu(n) \sum_{d \le x/n} LL(d)$$

and by Lemma 41.2 this is

$$= \sum_{n \le x} \mu(n)\left\{\frac{x}{n} \log^2 \frac{x}{n} + O\left(\frac{x}{n} \log \frac{x}{n}\right)\right\}$$

$$= x \sum_{n \le x} \frac{\mu(n)}{n} \log^2 \frac{x}{n} + O\left(x \sum_{n \le x} \frac{\mu(n)}{n} \log \frac{x}{n}\right)$$

$$= x\{2 \log x + O(1)\} + O(xO(1)) = 2x \log x + O(x).$$

EXERCISES

41-1. Prove Lemma 41.3.

41-2. Supply the details for the proof of Lemma 41.5.

41-3. Prove

$$\sum_{n \le x} \log \frac{x}{n} = x + 0(\log x).$$

42. The Function $\psi(x)$

We now introduce a technique from the calculus of finite differences which we will employ frequently in what follows. This method is known as "summation by parts" and, like integration by parts, it interchanges the roles of functions under a summation (integration) sign. The technique is concerned with sums of the type

(5) $f(1)g(1) + \{f(2) - f(1)\} g(2) + \cdots + \{f(m) - f(m-1)\} g(m).$

To simplify notation we do not hesitate to write the above sum as

(6) $$\sum_{n \le m} \{f(n) - f(n-1)\} g(n)$$

even though $f(0)$ may not be zero, or may not even be defined. Whenever we write (5) in the form of (6) it will be understood that we are using the convention that the term $f(0) g(1)$ is zero.

Theorem 42.1. (*Summation by parts.*) *Suppose f, g are functions of a real variable, that both are defined for all $x \ge 1$, and that the above convention is observed. Then*

$$\sum_{n \le x} \{f(n) - f(n-1)\}g(n) = f([x]) g([x]) + \sum_{n \le x-1} \{g(n) - g(n+1)\} f(n).$$

Proof. We have

$$\sum_{n \le x} \{f(n) - f(n-1)\}g(n)$$

$$= \sum_{n \le x} f(n) g(n) - \sum_{n \le x} f(n-1) g(n)$$

$$= f([x]) g([x]) + \sum_{n \le x-1} f(n) g(n) - \sum_{2 \le n \le x} f(n-1) g(n)$$

$$= f([x]) g([x]) + \sum_{n \le x-1} f(n) g(n) - \sum_{n \le x-1} f(n) g(n+1)$$

$$= f([x]) g([x]) + \sum_{n \le x-1} \{g(n) - g(n+1)\} f(n).$$

Some special cases of this theorem are very familiar. For example, if $g(n) = \iota_0(n) = 1$, then we obtain this well-known result on telescopic series:

$$\sum_{n \leq x} \{f(n) - f(n-1)\} = f([x]), \qquad \text{if } f(0) = 0.$$

By taking $f(n) = \iota_1(n) = n$, we have

(7) $$\sum_{n \leq x-1} \{g(n) - g(n+1)\}n = -[x]g([x]) + \sum_{n \leq x} g(n).$$

We have already used this special case of Theorem 42.1 in the proof of Theorem 20.1.

We proceed now to find some important formulas involving the function $\psi(x)$. We have

$$\psi(x) = \sum_{n \leq x} \Lambda(n) = \sum_{n \leq x} \mu \cdot L(n)$$

$$= \sum_{n \leq x} \mu(n) \sum_{d \leq x/n} L(d) = \sum_{n \leq x} \mu(n) 0\left(\frac{x}{n} \log \frac{x}{n}\right)$$

$$= 0\left(x \sum_{n \leq x} \frac{\mu(n)}{n} \log \frac{x}{n}\right) = 0(x)$$

so that

(8) $$\psi(x) = 0(x).$$

Lemma 42.1.

$$\sum_{n \leq x} \Lambda L(n) = \psi(x) \log x + 0(x)$$

Proof. Since $\psi(n) - \psi(n-1) = \Lambda(n)$, we use Theorem 42.1 to find

$$\sum_{n \leq x} \Lambda L(n) = \sum_{n \leq x} \{\psi(n) - \psi(n-1)\} L(n)$$

$$= \psi(x) \log [x] + \sum_{n \leq x-1} \{\log n - \log(n+1)\} \psi(n)$$

$$= \psi(x)\left\{\log x + 0\left(\frac{1}{x}\right)\right\} - \sum_{n \leq x-1} \log\left(1 + \frac{1}{n}\right) \psi(n)$$

$$= \psi(x) \log x + 0(\psi(x)) 0\left(\frac{1}{x}\right) + \sum_{n \leq x-1} 0\left(\frac{1}{n}\right) 0(n)$$

$$= \psi(x) \log x + 0(1) + 0(x).$$

This completes the proof.

Since

$$\sum_{n \leq x} L(n) = \sum_{n \leq x} \iota_0 \cdot \Lambda(n) = \sum_{n \leq x} \iota_0(n) \sum_{d \leq x/n} \Lambda(d) = \sum_{n \leq x} \psi\left(\frac{x}{n}\right)$$

we use Lemma 41.3 to conclude that

(9) $$\sum_{n \leq x} \psi\left(\frac{x}{n}\right) = x \log x - x + 0(\log x).$$

Lemma 42.2. *Let* $\Lambda_1(n) = \Lambda(n)/n$. *Then*

$$\sum_{n \leq x} \Lambda_1(n) = \log x + 0(1).$$

Proof. From (9), and its derivation,

$$x \log x + 0(x) = \sum_{n \leq x} \Lambda \cdot \iota_0(n) = \sum_{n \leq x} \Lambda(n) \sum_{d \leq x/n} 1$$

$$= \sum_{n \leq x} \Lambda(n) \left\{ \frac{x}{n} + 0(1) \right\}$$

$$= x \sum_{n \leq x} \Lambda_1(n) + 0\left(\sum_{n \leq x} \Lambda(n) \right)$$

$$= x \sum_{n \leq x} \Lambda_1(n) + 0(x),$$

from (8). Divide through by x to complete the argument.

Theorem 42.2.
$$\sum_{n \leq x} \Lambda_1 L(n) = \tfrac{1}{2} \log^2 x + 0(\log x)$$

Proof. Let

$$f(x) = \sum_{n \leq x} \Lambda_1(n) = \log x + 0(1).$$

Then

$$\sum_{n \leq x} \Lambda_1 L(n) = \sum_{n \leq x} \{ f(n) - f(n-1) \} L(n)$$

$$= f[(x)] \log [x] - \sum_{n \leq x-1} f(n) \log \left(1 + \frac{1}{n} \right)$$

$$= \left\{ \log x + 0\left(\frac{1}{x} \right) + 0(1) \right\} \left\{ \log x + 0\left(\frac{1}{x} \right) \right\}$$

$$- \sum_{n \leq x-1} \{ \log n + 0(1) \} \log \left(1 + \frac{1}{n} \right)$$

$$= \log^2 x + 0(\log x) - \sum_{n \leq x - 1} (\log n) \log \left(1 + \frac{1}{n} \right)$$

$$+ 0 \left(\sum_{n \leq x - 1} \log \left(1 + \frac{1}{n} \right) \right)$$

(10) $\sum_{n \leq x} \Lambda_1 L(n) = \log^2 x + 0(\log x) - S_2 + 0(S_3).$

Consider a McLaurin series expansion of $\log (1 + t)$ with $t = 1/n$:

$$\log \left(1 + \frac{1}{n} \right) = t + \sum_{k=2}^{\infty} \frac{(-1)^{k-1} t^k}{k} = t + 0 \left(\sum_{k=2}^{\infty} t^k \right) = t + 0 \left(t^2 \frac{1}{1-t} \right)$$

$$= \frac{1}{n} + 0(1/n^2).$$

Therefore,

$$S_2 = \sum_{n \leq x - 1} (\log n) \left(\frac{1}{n} + 0(1/n^2) \right)$$

$$= \sum_{n \leq x - 1} \frac{\log n}{n} + 0 \left(\sum_{n \leq x - 1} \frac{\log n}{n^2} \right)$$

$$= \tfrac{1}{2} \log^2 (x - 1) + C_3 + 0 \left(\frac{\log (x - 1)}{x - 1} \right) + 0 \left(\sum_{n=1}^{\infty} \frac{\log n}{n^2} \right)$$

$$= \tfrac{1}{2} \left\{ \log x \left(1 - \frac{1}{x} \right) \right\}^2 + C_3 + 0 \left(\frac{\log x}{x} \right) + 0(1)$$

since the infinite series converges. But

$$\left\{ \log x + \log \left(1 - \frac{1}{x} \right) \right\}^2 = \left\{ \log x + 0 \left(\frac{1}{x} \right) \right\}^2$$

$$= \log^2 x + 0 \left(\frac{\log x}{x} \right)$$

so that

$$S_2 = \tfrac{1}{2} \log^2 x + 0(1).$$

Also,

$$S_3 = \sum_{n \leq x - 1} \log \left(1 + \frac{1}{n} \right) = \sum_{n \leq x - 1} \{ \log (n + 1) - \log n \} = \log [x] = 0(\log x).$$

Making appropriate substitutions into (10), we complete the proof.

Corollary 42.2a.

$$\sum_{n \le x} \Lambda_1(n) \log \frac{x}{n} = \tfrac{1}{2} \log^2 x + 0(\log x)$$

Proof. Since $\log(x/n) = \log x - \log n$,

$$\sum_{n \le x} \Lambda_1(n) \log \frac{x}{n} = \log x \sum_{n \le x} \Lambda_1(n) - \sum_{n \le x} \Lambda_1 L(n)$$

$$= (\log x)\{\log x + 0(1)\} - \{\tfrac{1}{2} \log^2 x + 0(\log x)\}$$

from Theorem 42.2 and Lemma 42.2.

Corollary 42.2b.

$$\sum_{n \le x} \Lambda_1 \cdot \Lambda_1(n) = \tfrac{1}{2} \log^2 x + 0(\log x)$$

Proof.

$$\sum_{n \le x} \Lambda_1 \cdot \Lambda_1(n) = \sum_{n \le x} \Lambda_1(n) \sum_{d \le x/n} \Lambda_1(d)$$

$$= \sum_{n \le x} \Lambda_1(n) \left\{ \log \frac{x}{n} + 0(1) \right\}$$

$$= \sum_{n \le x} \Lambda_1(n) \log \frac{x}{n} + 0\left(\sum_{n \le x} \Lambda_1(n) \right)$$

$$= \tfrac{1}{2} \log^2 x + 0(\log x) + 0(\log x)$$

Theorem 42.3.

$$\sum_{n \le x} \Lambda \cdot (\Lambda L)(n) = \tfrac{1}{2} \log x \sum_{n \le x} \Lambda \cdot \Lambda(n) + 0(x \log x)$$

Proof. Let

$$S_4 = \sum_{n \le x} \Lambda \cdot (\Lambda L)(n).$$

Then

$$S_4 = \sum_{n \le x} \Lambda(n) \sum_{d \le x/n} \Lambda L(d)$$

$$= \sum_{n \le x} \Lambda(n) \left\{ \psi\left(\frac{x}{n}\right) \log \frac{x}{n} + 0\left(\frac{x}{n}\right) \right\}, \quad \text{from Lemma 42.1,}$$

$$= \sum_{n \le x} \Lambda(n) \psi\left(\frac{x}{n}\right) \log \frac{x}{n} + 0\left(x \sum_{n \le x} \Lambda_1(n) \right)$$

$$= \log x \sum_{n \leq x} \Lambda(n) \, \psi\left(\frac{x}{n}\right) - \sum_{n \leq x} \Lambda L(n) \, \psi\left(\frac{x}{n}\right) + 0(x \log x)$$

$$= \log x \sum_{n \leq x} \Lambda(n) \sum_{d \leq x/n} \Lambda(d) - \sum_{n \leq x} \Lambda L(n) \sum_{d \leq x/n} \Lambda(d) + 0(x \log x)$$

$$S_4 = \log x \sum_{n \leq x} \Lambda \cdot \Lambda(n) - S_4 + 0(x \log x).$$

The proof is completed by solving the last equation for S_4.

Theorem 42.4.

$$\sum_{n \leq x} \Lambda \cdot (\Lambda L)(n) + \sum_{n \leq x} \Lambda \cdot \Lambda \cdot \Lambda(n) = x \log^2 x + 0(x \log x)$$

Proof. From Theorem 41.1 we have

$$\sum_{d \leq x/n} \Lambda L(d) + \sum_{d \leq x/n} \Lambda \cdot \Lambda(d) = \frac{2x}{n} \log \frac{x}{n} + 0\left(\frac{x}{n}\right).$$

We multiply each term by $\Lambda(n)$ and then sum over $n \leq x$ to get

$$\sum_{n \leq x} \Lambda(n) \sum_{d \leq x/n} \Lambda L(d) + \sum_{n \leq x} \Lambda(n) \sum_{d \leq x/n} \Lambda \cdot \Lambda(d) = 2x \sum_{n \leq x} \Lambda_1(n) \log \frac{x}{n}$$

$$+ 0\left(x \sum_{n \leq x} \Lambda_1(n)\right)$$

$$\sum_{n \leq x} \Lambda \cdot (\Lambda L)(n) + \sum_{n \leq x} \Lambda \cdot \Lambda \cdot \Lambda(n) = 2x\{\tfrac{1}{2} \log^2 x + 0(\log x)\}$$

$$+ 0(x \log x)$$

from Corollary 42.2a and Lemma 42.2. The right side of the last equation is obviously $x \log^2 x + 0(x \log x)$.

EXERCISES

42-1.   Use Theorem 42.3 and 42.4 to prove that

$$\log x \sum_{n \leq x} \Lambda \cdot \Lambda(n) + 2 \sum_{n \leq x} \Lambda \cdot \Lambda \cdot \Lambda(n) = 2x \log^2 x + 0(x \log x).$$

42-2.* Use Lemma 42.1 and Theorem 41.1 to prove *Selberg's formula*:

(11) $$\psi(x) \log x + \sum_{n \leq x} \Lambda \cdot \Lambda(n) = 2x \log x + 0(x).$$

Hence prove that

$$\psi(x) \log^2 x + \log x \sum_{n \leq x} \Lambda \cdot \Lambda(n) = 2x \log^2 x + 0(x \log x).$$

42-3.* Combine the results of the two previous exercises to prove

$$(12) \qquad \psi(x) \log^2 x = 2 \sum_{n \leq x} \Lambda \cdot \Lambda \cdot \Lambda(n) + 0(x \log x).$$

42-4.* From Exercise 42.2 and Theorem 42.3 prove

$$(13) \qquad \psi(x) \log^2 x + 2 \sum_{n \leq x} \Lambda \cdot (\Lambda L)(n) = 2x \log^2 x + 0(x \log x).$$

42-5.* Prove

$$\frac{\Lambda \cdot \Lambda(n)}{n} = \Lambda_1 \cdot \Lambda_1(n)$$

for all $n \in \mathscr{Z}^+$.

43. A Fundamental Inequality

We now combine the results of the previous section to obtain a basic inequality which will be used to estimate the size of the quotient $\psi(x)/x$. Our work is simplified somewhat if we introduce the function $r(x)$ defined by

$$r(x) = \psi(x) - x.$$

We will also define the functions α, β, γ as follows:

$$\alpha(x) = \sum_{n \leq x} \Lambda L(n),$$

$$\beta(x) = \sum_{n \leq x} \Lambda \cdot \Lambda(n),$$

$$\gamma(x) = (\alpha + \beta)(x).$$

Lemma 43.1.

$$|r(x)| \log^2 x \leq 2 \sum_{n \leq x} \Lambda \cdot \Lambda(n) \left| r\left(\frac{x}{n}\right) \right| + 0(x \log x)$$

Proof. In (12) replace $\psi(x)$ by $r(x) + x$. Then

$$r(x) \log^2 x + x \log^2 x = 2 \sum_{n \leq x} (\Lambda \cdot \Lambda) \cdot \Lambda(n) + 0(x \log x)$$

$$= 2 \sum_{n \leq x} \Lambda \cdot \Lambda(n) \sum_{d \leq x/n} \Lambda(d) + 0(x \log x)$$

$$= 2 \sum_{n \leq x} \Lambda \cdot \Lambda(n) \, \psi\left(\frac{x}{n}\right) + 0(x \log x)$$

$$= 2 \sum_{n \leq x} \Lambda \cdot \Lambda(n) \left\{ r\left(\frac{x}{n}\right) + \frac{x}{n} \right\} + 0(x \log x)$$

$$= 2x \sum_{n \le x} \frac{\Lambda \cdot \Lambda(n)}{n} + 2 \sum_{n \le x} \Lambda \cdot \Lambda(n)\, r\left(\frac{x}{n}\right) + O(x \log x)$$

$$= 2x\{\tfrac{1}{2} \log^2 x + O(\log x)\} + 2 \sum_{n \le x} \Lambda \cdot \Lambda(n)\, r\left(\frac{x}{n}\right)$$

$$+ O(x \log x)$$

by Exercise 42-5 and Corollary 42.2b. Therefore,

$$r(x) \log^2 x = 2 \sum_{n \le x} \Lambda \cdot \Lambda(n)\, r\left(\frac{x}{n}\right) + O(x \log x).$$

Taking absolute values, we have

$$|r(x)| \log^2 x \le 2 \sum_{n \le x} \Lambda \cdot \Lambda(n) \left| r\left(\frac{x}{n}\right) \right| + O(x \log x)$$

since $\Lambda \cdot \Lambda(n) \ge 0$ for all n.

Lemma 43.2.

$$|r(x)| \log^2 x \le 2 \sum_{n \le x} \Lambda L(n) \left| r\left(\frac{x}{n}\right) \right| + O(x \log x)$$

Proof. We first notice that

$$\sum_{n \le x} (\Lambda L) \cdot \Lambda(n) = \sum_{n \le x} \Lambda L(n) \sum_{d \le x/n} \Lambda(d)$$

$$= \sum_{n \le x} \Lambda L(n)\, \psi\left(\frac{x}{n}\right)$$

$$= \sum_{n \le x} \Lambda L(n) \left\{ r\left(\frac{x}{n}\right) + \frac{x}{n} \right\}$$

$$= \sum_{n \le x} \Lambda L(n)\, r\left(\frac{x}{n}\right) + x\{\tfrac{1}{2} \log^2 x + O(\log x)\}$$

by Theorem 42.2. Now in (13) replace $\psi(x)$ by $r(x) + x$ to get

$$r(x) \log^2 x = -2 \sum_{n \le x} (\Lambda L) \cdot \Lambda(n) + x \log^2 x + O(x \log x)$$

$$= -2 \sum_{n \le x} \Lambda L(n) \left\{ r\left(\frac{x}{n}\right) + \frac{x}{n} \right\} + x \log^2 x + O(x \log x)$$

$$= -2 \sum_{n \le x} \Lambda L(n)\, r\left(\frac{x}{n}\right) - 2x\{\tfrac{1}{2} \log^2 x + O(\log x)\}$$

$$+ x \log^2 x + O(x \log x),$$

by Theorem 42.2; therefore,

$$r(x) \log^2 x = -2 \sum_{n \le x} \Lambda L(n) \, r\left(\frac{x}{n}\right) + 0(x \log x).$$

Again the proof is completed by taking absolute values and using the triangle inequality.

Theorem 43.1.

$$|r(x)| \log^2 x \le \sum_{n \le x} (\log n) \left| r\left(\frac{x}{n}\right) \right| + 0(x \log x)$$

Proof. By adding corresponding members of the inequalities of the last two lemmas, we have

(14) $$|r(x)| \log^2 x \le \sum_{n \le x} ((\Lambda L) + (\Lambda \cdot \Lambda))(n) \left| r\left(\frac{x}{n}\right) \right| + 0(x \log x).$$

Since $\Lambda L(n) = \alpha(n) - \alpha(n-1)$ and $\Lambda \cdot \Lambda(n) = \beta(n) - \beta(n-1)$, and $\gamma(n) = 2n \log n + 0(n)$ by Theorem 41.1, we get

$$\sum_{n \le x} ((\Lambda L) + (\Lambda \cdot \Lambda))(n) \left| r\left(\frac{x}{n}\right) \right|$$

$$= \sum_{n \le x} \{\alpha(n) + \beta(n) - \alpha(n-1) - \beta(n-1)\} \left| r\left(\frac{x}{n}\right) \right|$$

$$= \sum_{n \le x} \{\gamma(n) - \gamma(n-1)\} \left| r\left(\frac{x}{n}\right) \right|$$

$$= \gamma([x]) \left| r\left(\frac{x}{[x]}\right) \right| + \sum_{n \le x-1} \left\{ \left| r\left(\frac{x}{n}\right) \right| - \left| r\left(\frac{x}{n+1}\right) \right| \right\} \gamma(n)$$

$$= \gamma(x) \, 0(1) + \sum_{n \le x-1} \left\{ \left| r\left(\frac{x}{n}\right) \right| - \left| r\left(\frac{x}{n+1}\right) \right| \right\} \{2n \log n + 0(n)\}$$

$$= 0(x \log x) + 2 \sum_{n \le x-1} n(\log n) \left\{ \left| r\left(\frac{x}{n}\right) \right| - \left| r\left(\frac{x}{n+1}\right) \right| \right\}$$

$$+ 0\left(\sum_{n \le x-1} n \left\{ \left| r\left(\frac{x}{n}\right) \right| - \left| r\left(\frac{x}{n+1}\right) \right| \right\} \right)$$

(15) $$= 0(x \log x) + S_5 + 0(S_6).$$

Now we apply summation by parts on S_5 and get

$$S_5 = -2[x](\log [x]) \left| r\left(\frac{x}{[x]}\right) \right| + 2 \sum_{n \le x} \{n \log n - (n-1) \log (n-1)\} \left| r\left(\frac{x}{n}\right) \right|.$$

But

$$[x](\log [x]) \left| r\left(\frac{x}{[x]}\right)\right| = \{x + 0(1)\}\left\{\log x + 0\left(\frac{1}{x}\right)\right\}0(1) = 0(x \log x)$$

and

$$n \log n - (n - 1) \log (n - 1) = \log \frac{n^n}{(n - 1)^{n-1}}$$

$$= \log n\left(\frac{n}{n - 1}\right)^{n-1}$$

$$= \log n + \log \left(\frac{n}{n - 1}\right)^{n-1}$$

$$= \log n + 0(1)$$

since $\log (n/(n - 1))^{n-1} \to 1$ as $n \to \infty$. Therefore,

$$S_5 = 0(x \log x) + 2 \sum_{n \leq x} (\log n)\left| r\left(\frac{x}{n}\right)\right| + 0\left(\sum_{n \leq x}\left| r\left(\frac{x}{n}\right)\right|\right)$$

$$= 2 \sum_{n \leq x} (\log n)\left| r\left(\frac{x}{n}\right)\right| + 0(x \log x),$$

because

$$\sum\left| r\left(\frac{x}{n}\right)\right| \leq \sum \psi\left(\frac{x}{n}\right) + \sum \frac{x}{n} + 0(x \log x)$$

by (9).

Furthermore, since $|u| - |v| \leq |u - v|$,

$$S_6 \leq \sum_{n \leq x-1} n\left| r\left(\frac{x}{n}\right) - r\left(\frac{x}{n + 1}\right)\right|$$

$$= \sum_{n \leq x-1} n\left| \psi\left(\frac{x}{n}\right) - \psi\left(\frac{x}{n + 1}\right) - \left(\frac{x}{n} - \frac{x}{n + 1}\right)\right|$$

$$\leq \sum_{n \leq x-1} n\left\{\left| \psi\left(\frac{x}{n}\right) - \psi\left(\frac{x}{n + 1}\right)\right| + \left| \frac{x}{n} - \frac{x}{n + 1}\right|\right\}.$$

Because

$$\psi\left(\frac{x}{n}\right) - \psi\left(\frac{x}{n + 1}\right) \geq 0 \quad \text{and} \quad \frac{x}{n} - \frac{x}{n + 1} > 0$$

we may remove the absolute value signs and write

$$S_6 \leq \sum_{n \leq x-1} n\left\{\psi\left(\frac{x}{n}\right) - \psi\left(\frac{x}{n+1}\right)\right\} + \sum_{n \leq x-1} n\left\{\frac{x}{n} - \frac{x}{n+1}\right\}$$

$$= -[x]\psi\left(\frac{x}{[x]}\right) + \sum_{n \leq x} \psi\left(\frac{x}{n}\right) + \sum_{n \leq x-1} \frac{x}{n+1}, \quad \text{by (7)}$$

$$= 0 + 0(x \log x) + 0(x \log x)$$

so that $S_6 = 0(x \log x)$. Making substitutions for the values of S_5 and S_6 into (15) we have

$$\sum_{n \leq x} ((\Lambda L) + (\Lambda \cdot \Lambda))(n)\left|r\left(\frac{x}{n}\right)\right| = 2 \sum_{n \leq x} (\log n)\left|r\left(\frac{x}{n}\right)\right| + 0(x \log x).$$

This into (14) proves the theorem.

44. The Behavior of $\dfrac{r(x)}{x}$

We will show in this section that $r(x) = o(x)$, or equivalently, that $r(x)/x \to 0$ as $x \to \infty$. Our procedure will be first to show that there are infinitely many values n for which $r(n)/n$ is small. Then we will prove that corresponding to each such n, there is an interval (containing n) throughout which the quotient $r(x)/x$ stays small in absolute value. Finally, we will extend these results to show that $r(x) = o(x)$.

Lemma 44.1. *Let $k > 1$ be an arbitrary, fixed real number. There exists a positive constant C_4 (independent of k) and a number x_0 (depending only on k) such that for every $x \geq x_0$ there is an integer $N \in [x, kx]$ for which*

$$\left|\frac{r(N)}{N}\right| < \frac{C_4}{\log k}.$$

Proof. In Lemma 42.2, we proved $\sum \Lambda_1(n) = \log x + 0(1)$. But

$$\sum_{n \leq x} \Lambda_1(n) = \sum_{n \leq x} \Lambda(n)\frac{1}{n} = \sum_{n \leq x} \{\psi(n) - \psi(n-1)\}\frac{1}{n}$$

$$= \psi([x])\frac{1}{[x]} + \sum_{n \leq x-1} \left\{\frac{1}{n} - \frac{1}{n+1}\right\}\psi(n)$$

$$= \psi(x)\, 0\left(\frac{1}{x}\right) + \sum_{n \leq x-1} \frac{\psi(n)}{n(n+1)}$$

$$= 0(1) + \sum_{n \leq x-1} \frac{n + r(n)}{n(n+1)}$$

$$= 0(1) + \sum_{n \le x-1} \frac{1}{n+1} + \sum_{n \le x-1} \frac{r(n)}{n(n+1)}$$

$$= 0(1) + \{\log x + 0(1)\} + \sum_{n \le x-1} \frac{r(n)}{n^2} \left\{ 1 - \frac{1}{n+1} \right\}$$

$$= \log x + 0(1) + \sum_{n \le x-1} \frac{r(n)}{n^2} - \sum_{n \le x-1} \frac{r(n)}{n^2(n+1)}.$$

But $r(n) = \psi(n) - n = 0(n)$, so

$$\sum_{n \le x-1} \frac{r(n)}{n^2(n+1)} = 0 \left(\sum_{n \le x-1} \frac{n}{n^2(n+1)} \right) = 0 \left(\sum_{n=1}^{\infty} \frac{1}{n(n+1)} \right) = 0(1)$$

and

$$\sum_{n \le x-1} \frac{r(n)}{n^2} = \sum_{n \le x} \frac{r(n)}{n^2} + 0(1).$$

Therefore,

$$\log x + 0(1) + \sum_{n \le x} \frac{r(n)}{n^2} = \sum_{n \le x} \Lambda_1(n) = \log x + 0(1)$$

and it follows that

$$\sum_{n \le x} \frac{r(n)}{n^2} = 0(1).$$

Thus, there is a constant C such that

$$\left| \sum_{n \le x} \frac{r(n)}{n^2} \right| < C \quad \text{for all } x.$$

Hence, if $\mathcal{I} = (x, kx]$,

$$\sum_{n \in \mathcal{I}} \frac{r(n)}{n^2} = \sum_{n \le kx} \frac{r(n)}{n^2} - \sum_{n \le x} \frac{r(n)}{n^2}$$

so

$$\left| \sum_{n \in \mathcal{I}} \frac{r(n)}{n^2} \right| \le \left| \sum_{n \le kx} \frac{r(n)}{n^2} \right| + \left| \sum_{n \le x} \frac{r(n)}{n^2} \right| < 2C.$$

We now argue differently depending on whether $r(n)$ does or does not change sign on the interval $(x, kx]$.

Case 1, r(n) does not change sign for $n \in \mathcal{I}$.

Then

$$\left| \sum_{n \in \mathcal{I}} \frac{r(n)}{n^2} \right| = \sum_{n \in \mathcal{I}} \left| \frac{r(n)}{n^2} \right| < 2C.$$

Let

$$\left| \frac{r(N)}{N} \right| = \min_{n \in \mathcal{I}} \left| \frac{r(n)}{n} \right|.$$

Hence,

$$2C > \sum_{n \in \mathcal{I}} \left| \frac{r(n)}{n^2} \right| = \sum_{n \in \mathcal{I}} \left| \frac{r(n)}{n} \right| \frac{1}{n} \geq \left| \frac{r(N)}{N} \right| \sum_{n \in \mathcal{I}} \frac{1}{n}$$

$$= \left| \frac{r(N)}{N} \right| \left\{ \sum_{n \leq kx} \frac{1}{n} - \sum_{n \leq x} \frac{1}{n} \right\}$$

$$= \left| \frac{r(N)}{N} \right| \left\{ \log kx + \gamma + 0\left(\frac{1}{x}\right) - \log x - \gamma + 0\left(\frac{1}{x}\right) \right\}$$

$$= \left| \frac{r(N)}{N} \right| \left\{ \log k + 0\left(\frac{1}{x}\right) \right\}.$$

Thus we have

$$2C > \left| \frac{r(N)}{N} \right| \left\{ \log k + 0\left(\frac{1}{x}\right) \right\}.$$

Since $k > 1$, $\frac{1}{2} \log k > 0$; since $0(1/x) \to 0$ as $x \to \infty$, there exists x_0' depending only on k such that if $x \geq x_0'$, then $0(1/x) \geq -\frac{1}{2} \log k$. Therefore, if $x \geq x_0'$, $\log k + 0(1/x) \geq \frac{1}{2} \log k > 0$, and

$$\left| \frac{r(N)}{N} \right| < \frac{2C}{\frac{1}{2} \log k} = \frac{4C}{\log k}.$$

In this case, then, we may take $C_4 \geq 4C$.

Case 2, r(n) changes sign on the interval $\mathcal{I}$.

Suppose one sign change occurs at $n = N$. Then

$$|r(N)| < |r(N) - r(N-1)| = |\psi(N) - \psi(N-1) - (N - (N-1))|$$

$$\leq \psi(N) - \psi(N-1) + 1 = \Lambda(N) + 1 \leq \log N + 1.$$

Therefore,

$$\left| \frac{r(N)}{N} \right| < \frac{\log N + 1}{N}.$$

But since $N \in \mathscr{J}, x < N$, and the function $(\log y + 1)/y$ is a decreasing function (the derivative is negative) for $y > 1$, we have

$$\frac{\log N + 1}{N} < \frac{\log x + 1}{x}$$

and

$$\left| \frac{r(N)}{N} \right| < \frac{\log x + 1}{x}.$$

Now, $(\log x + 1)/x \to 0$ as $x \to \infty$, so there exists x_0'' depending only on k such that if $x \geq x_0''$, then $(\log x + 1)/x < 1/(\log k)$. Hence for $x \geq x_0''$,

$$\left| \frac{r(N)}{N} \right| < \frac{1}{\log k}.$$

and here we could take $C_4 \geq 1$.

We incorporate the results of the two cases into our choices for x_0 and C_4. For reasons which will become clear in the proof of the next theorem, we also want $C_4 \geq \log^2 4$. Therefore, choose

$$x_0 = \max \{x_0', x_0''\}$$

$$C_4 = \max \{4C, \log^2 4\}$$

and the proof is complete.

Notice that Lemma 44.1 implies there are infinitely many integers n such that $|r(n)/n|$ is arbitrarily small. For example, we could argue in this way. Let ε be a positive number. Since C_4 does not depend on k, we can take k so large that $C_4/\log k < \varepsilon$, so that when $x \geq x_0$, there is an $N_1 \in [x,kx]$ such that $|r(N_1)/N_1| < \varepsilon$. Now apply the lemma again with $x = N_1 + 1$: there exists $N_2 \in [N_1 + 1,k(N_1 + 1)]$ such that $|r(N_2)/N_2| < \varepsilon$, and obviously $N_2 \neq N_1$. Continue by induction.

Theorem 44.1. *Let δ be any positive number such that $\delta < 8 \log 2$. Then there exists x_0 (depending only on δ) such that if $x \geq x_0$, then the interval $[x,xe^{C_4/\delta}]$ contains a subinterval $\mathscr{J} = [N,Ne^{\delta/8}]$ such that for all $y \in \mathscr{J}$,*

$$\left| \frac{r(y)}{y} \right| < \delta.$$

Proof. In Lemma 44.1, take $k = \frac{1}{2}e^{4C_4/\delta}$. Then the x_0 of that lemma is completely determined by δ, call it x_1'; if $x \geq x_1'$, then there exists $N \in [x,\frac{1}{2}xe^{4C_4/\delta}]$ such that

$$\left| \frac{r(N)}{N} \right| < \frac{C_4}{\log \frac{1}{2}e^{4C_4/\delta}}.$$

But

$$\log \tfrac{1}{2}e^{4C_4/\delta} = \log \tfrac{1}{2} + \frac{4C_4}{\delta} = \frac{4C_4}{\delta} - \log 2$$

and $\log 2 \le 2C_4/\delta$, because if not, then

$$\log 2 > \frac{2C_4}{\delta} \ge \frac{2(\log 4)^2}{8 \log 2} = \log 2.$$

Therefore,

$$\frac{4C_4}{\delta} - \log 2 \ge \frac{2C_4}{\delta}$$

and

$$\left| \frac{r(N)}{N} \right| < \frac{C_4}{4C_4/\delta - \log 2} \le \frac{C_4}{2C_4/\delta} = \frac{\delta}{2}.$$

We show that this N is the N in the theorem. First, $x \le N$ is obvious, and

$$N \le \tfrac{1}{2}e^{4C_4/\delta}x.$$

But

$$e^{\delta/8} < e^{(8 \log 2)/8} = 2$$

so

$$e^{\delta/8}N < 2N \le e^{4C_4/\delta}x.$$

This proves that $\mathscr{J} = [N, e^{\delta/8}N] \subset [x, e^{4C_4/\delta}x]$, and it remains only to show that for all $y \in \mathscr{J}$, $|r(y)/y| < \delta$.

Suppose $y \in \mathscr{J}$. In (11), let $x = y$ and then $x = N$, and subtract the two formulas to get

$$\psi(y) \log y - \psi(N) \log N + \sum_{n \in \mathscr{J}_1} \Lambda \cdot \Lambda(n) = 2y \log y - 2N \log N + 0(y) + 0(N)$$

where $\mathscr{J}_1$ is the interval $(N, y]$. Since $y \le e^{\delta/8}N < 2N$, $0(y) = 0(N)$; since $\Lambda \cdot \Lambda(n) \ge 0$, the above summation is ≥ 0, and we have

$$\psi(y) \log y - \psi(N) \log N \le 2y \log y - 2N \log N + 0(N).$$

We rewrite this in the form

$$\{\psi(y) - \psi(N)\} \log y + \psi(N) \log \frac{y}{N} \le 2(y - N) \log y + 2N \log \frac{y}{N} + 0(N).$$

Now, $y/N < 2$, so $\log(y/N) = 0(1)$, and $\psi(N) = 0(N)$. We use these observations above and get

$$\{\psi(y) - \psi(N)\} \log y \le 2(y - N) \log y + 0(N).$$

If we divide through by log y, we have the error term $0(N/\log y) = 0(N/\log N)$, because $N < y$, so

(16)
$$\psi(y) - \psi(N) \leq 2(y - N) + 0\left(\frac{N}{\log N}\right).$$

But

$$|r(y)| - |r(N)| - (y - N) \leq |r(y) - r(N)| - (y - N)$$
$$\leq |r(y) - r(N) - (y - N)|$$
$$= |\psi(y) - \psi(N)| = \psi(y) - \psi(N),$$

so from (16) we have

$$|r(y)| \leq |r(N)| + 3(y - N) + 0\left(\frac{N}{\log N}\right)$$

(17)
$$\left|\frac{r(y)}{y}\right| \leq \left|\frac{r(N)}{y}\right| + \frac{3(y - N)}{y} + 0\left(\frac{N/y}{\log N}\right).$$

The error term here is $0(1/\log x)$ because $N/y < 1$ and $x \leq N$ implies

$$\frac{N/y}{\log N} < \frac{1}{\log N} \leq \frac{1}{\log x}.$$

Since $0(1/\log x) \to 0$ as $x \to \infty$, there exists x_1'' (depending only on δ) such that if $x \geq x_1''$, then $0(1/\log x) < \delta/8$. Choose

$$x_0 = \max \{x_1', x_1''\}.$$

Then for $x \geq x_0$,

$$\left|\frac{r(N)}{y}\right| = \frac{N}{y}\left|\frac{r(N)}{N}\right| < \frac{\delta}{2}$$

and

$$0\left(\frac{N/y}{\log N}\right) = 0\left(\frac{1}{\log x}\right) < \frac{\delta}{8}.$$

Since $y \leq e^{\delta/8}N$, $N/y \geq e^{-\delta/8}$. We would like to show this is $\geq 1 - \delta/8$. This is easy to see because

$$e^{-\delta/8} = \sum_{n=0}^{\infty} \frac{(-\delta/8)^n}{n!} = 1 - \frac{\delta}{8} + \sum_{n=2}^{\infty} \frac{(-\delta/8)^n}{n!} > 1 - \frac{\delta}{8}.$$

Therefore,

$$\frac{3(y - N)}{y} = 3\left(1 - \frac{N}{y}\right) \leq 3\left\{1 - \left(1 - \frac{\delta}{8}\right)\right\} = \frac{3\delta}{8}.$$

Using these estimates in (17), we have: if $x \geq x_0$, then

$$\left| \frac{r(y)}{y} \right| < \frac{\delta}{2} + \frac{3\delta}{8} + \frac{\delta}{8} = \delta \quad \text{for all } y \in \mathcal{J}.$$

The proof is complete.

Theorem 44.2.

$$r(x) = o(x)$$

Proof. We know $r(x) = O(x)$, so there is a constant C_5 such that $|r(x)| < C_5 x$ for all x. Using this in Theorem 43.1, we get

$$|r(x)| \log^2 x \leq 2 \sum_{n \leq x} (\log n) \frac{C_5 x}{n} + O(x \log x)$$

(18)

$$= 2C_5 x \sum_{n \leq x} \frac{\log n}{n} + O(x \log x).$$

Choose a number c such that $0 < c < 1$ and $cC_5 < 8 \log 2$, and then take $\delta = cC_5$. This determines the x_0 of Theorem 44.1. Now in (18), for those terms arising from $x/n \geq x_0$ and in one of the intervals $[N, e^{\delta/8} N]$, we could have gotten a better bound by using $|r(x/n)| < \delta x/n$ instead of $|r(x/n)| < C_5 x/n$, and by so doing we would have improved the estimate by a factor $(C_5 - \delta)$ for those terms where x/n is in one of the intervals.

Let $\theta = e^{4C_4/\delta}$ and $\lambda = e^{\delta/8}$. Then

$$\mathcal{J}_i = [\theta^{i-1}, \theta^i], \quad i = 1, 2, \ldots$$

is a sequence of intervals of the type in Theorem 44.1. If we choose a such that $\theta^a \leq x < \theta^{a+1}$, then $a \log \theta \leq \log x < (a + 1) \log \theta$, so

$$a = \left[\frac{\log x}{\log \theta} \right] = \left[\frac{\delta \log x}{4C_4} \right] = O(\log x).$$

If $\theta^b \geq x_0$, then each of the intervals $\mathcal{J}_i$ ($i = b + 1, \ldots, a$) contains a sub-interval $\mathcal{J}_i = [N_i, \lambda N_i]$ over which $|r(n)/n| < \delta$. We use the improved bound in (18) and get

$$|r(x)| \log^2 x \leq 2C_5 x \sum_{n \leq x} \frac{\log n}{n} - 2(C_5 - \delta)x \sum_{i=b+1}^{a} \sum_{x/n \in \mathcal{J}_i} \frac{\log n}{n} + O(x \log x).$$

(19)

To estimate the double summation occurring here, first consider

$$\sum_{x/n \in \mathscr{I}_i} \frac{\log n}{n} = \sum_{N_i \le x/n \le \lambda N_i} \frac{\log n}{n} \ge \sum_{x/\lambda N_i < n \le x/N_i} \frac{\log n}{n}$$

$$\ge \sum_{n \le x/N_i} \frac{\log n}{n} - \sum_{n \le x/\lambda N_i} \frac{\log n}{n}$$

$$= \left\{ \frac{1}{2} \log^2 \left(\frac{x}{N_i} \right) + C_3 + 0 \left(\frac{\log (x/N_i)}{x/N_i} \right) \right\}$$

$$- \left\{ \frac{1}{2} \log^2 \left(\frac{x}{\lambda N_i} \right) + C_3 + 0 \left(\frac{\log (x/\lambda N_i)}{x/\lambda N_i} \right) \right\}$$

by Lemma 41.4. Both of the error terms are $0(1)$, and

$$\frac{1}{2} \left\{ \left(\log \frac{x}{N} \right)^2 - \left(\log \frac{x}{\lambda N} \right)^2 \right\} = \frac{1}{2} \left\{ \left(\log \frac{x}{N} \right)^2 - \left(\log \frac{x}{N} - \log \lambda \right)^2 \right\}$$

$$= \frac{1}{2} \left\{ 2(\log \lambda) \log \frac{x}{N} - \log^2 \lambda \right\} = \frac{\delta}{8} \log \frac{x}{N} + 0(1).$$

Therefore,

$$\sum_{i=b+1}^{a} \sum_{x/n \in \mathscr{I}_i} \frac{\log n}{n} = \sum_{i=b+1}^{a} \left\{ \frac{\delta}{8} \log \frac{x}{N_i} + 0(1) \right\}$$

$$\ge \frac{\delta}{8} \sum_{i=b+1}^{a/2} \log \frac{x}{N_i} + 0 \left(\sum_{i=b+1}^{a} 1 \right)$$

(20) $$\ge \frac{\delta}{8} \sum_{i=b+1}^{a/2} \log \frac{x}{\theta^i} + 0(a).$$

In this last step, we have used the fact that

$$\mathscr{I}_i = [N, \lambda N_i] \subset \mathscr{I}_i = [\theta^{i-1}, \theta^i]$$

so $x/N_i > x/\theta^i$. Since $i \le a/2$, $\theta^i \le \theta^{a/2} \le x^{1/2}$, so $\log (x/\theta^i) \ge \log x^{1/2} = \frac{1}{2} \log x$. Also, $0(a) = 0(\log x)$ In (20), then,

$$\sum_{i=b+1}^{a} \sum_{x/n \in \mathscr{I}_i} \frac{\log n}{n} \ge \frac{\delta}{8} \sum_{i=1}^{a/2} \frac{1}{2} \log x - \frac{\delta}{8} \sum_{i=1}^{b} \frac{1}{2} \log x + 0(\log x)$$

$$= \frac{\delta}{16} (\log x) \frac{a}{2} - \frac{\delta}{16} (\log x) b + 0(\log x)$$

$$= \frac{\delta}{16}(\log x)\left\{\frac{1}{2}\frac{\delta \log x}{4C_4} + 0(1)\right\} - \frac{\delta}{16}(\log x)\left[\frac{\log x_0}{\log \theta}\right]$$

$$+ 0(\log x)$$

$$= \frac{\delta^2}{128}\frac{\log^2 x}{C_4} + 0(\log x)$$

since $\log x_0 = 0(1)$.

Putting this into (19), and using Lemma 41.4, we have

$$|r(x)| \log^2 x \le 2C_5 x \left\{\frac{1}{2}\log^2 x + C_3 + 0\left(\frac{\log x}{x}\right)\right\}$$

$$- 2(C_5 - \delta)x\left\{\frac{\delta^2 \log^2 x}{128C_4} + 0(\log x)\right\} + 0(x \log x)$$

$$\le \left\{C_5 - \frac{(C_5 - \delta)\delta^2}{64C_4}\right\}x \log^2 x + 0(x \log x)$$

so that

$$\left|\frac{r(x)}{x}\right| \le C_5 - \frac{(C_5 - \delta)\delta^2}{64C_4} + 0\left(\frac{1}{\log x}\right).$$

But $0(1/\log x) \to 0$ as $x \to \infty$, so there is some x_0' such that if $x \ge x_0'$, then

$$0\left(\frac{1}{\log x}\right) < \frac{(C_5 - \delta)\delta^2}{128C_4}$$

so for $x \ge x_0$ and $x \ge x_0'$, we have

$$\left|\frac{r(x)}{x}\right| \le C_5 - \frac{(C_5 - \delta)\delta^2}{128C_4}.$$

The bound here is

$$B_1 = C_5 - \frac{(C_5 - \delta)\delta^2}{128C_4} = C_5 - \frac{(C_5 c)^2 C_5(1 - c)}{128C_4} = C_5 - dC_5^3$$

where

$$d = \frac{(1 - c)c^2}{128C_4} > 0$$

so $B_1 < C_5$. Also, $B_1 > 0$, because

$$128C_4 C_5 - (C_5 - \delta)\delta^2 \ge 128(\log^2 4)C_5 - (C_5 - \delta)(64 \log^2 2)$$

$$= \{128(4\log^2 2) - 64 \log^2 2\}C_5 + 64\delta \log^2 2 > 0.$$

The important thing is that B_1 is a better bound than C_5. Now we can repeat the process starting with B_1; we can use the *same* c (since $cB_1 < cC_5 <$ 8 log 2 allows us to choose $\delta = cB_1$) and this will result in the *same* d (because d depends only on c and C_4, which is constant), and we will get

$$\left| \frac{r(x)}{x} \right| < B_1 - dB_1^3.$$

Call this $B_2 = B_1 - dB_1^3$, and we have $0 < B_2 < B_1$. Continuing by induction, we find a sequence $B_1 > B_2 > \cdots > 0$ which must converge because the sequence is decreasing and bounded. Suppose $B_i \to B$ as $i \to \infty$. We have the B_i satisfying $B_{i+1} = B_i - dB_i^3$. Taking limits as $i \to \infty$, we have $B = B - dB^3$, so $dB^3 = 0$. But $d > 0$, so $B = 0$. Hence, as $x \to \infty$, $|r(x)/x| \to 0$, and this is the statement of the theorem.

Corollary 44.2.

$$\psi(x) \sim x$$

Proof. We have $\psi(x) - x = r(x) = o(x)$. Therefore,

$$0 = \lim_{x \to \infty} \frac{\psi(x) - x}{x} = \lim_{x \to \infty} \frac{\psi(x)}{x} - 1, \quad \text{or} \quad 1 = \lim_{x \to \infty} \frac{\psi(x)}{x}.$$

Note. The argument used here applies in general, and the arguments are reversible. Thus $f(x) \sim g(x)$ if and only if $f(x) = g(x) + o(g(x))$.

EXERCISES

44-1.* Prove the statement made in the Note above. Also prove that $f(x) = g(x) + o(g(x))$ if and only if $f(x) = g(x) + o(f(x))$.

45. The Prime Number Theorem and Related Results

The Prime Number Theorem is equivalent to Theorem 44.2. We will first prove that Theorem 44.2 implies the Prime Number Theorem, and then we will give several equivalent statements.

Theorem 45.1. (*Prime Number Theorem.*)

$$\pi(x) \sim \frac{x}{\log x}$$

Proof. For each prime p, the number of terms log p in $\psi(x)$ is exactly a, where a satisfies $p^a \le x < p^{a+1}$. Thus, $a = [(\log x)/(\log p)]$, and

$$\psi(x) = \sum_{p^a \le x} \log p = \sum_{p \le x} \left[\frac{\log x}{\log p} \right] \log p \le \sum_{p \le x} \log x = \pi(x) \log x$$

so we have

$$\frac{\psi(x)}{x} \le \frac{\pi(x) \log x}{x}.$$

Now let $b = 1 - 1/\log \log x$. Then $x^b < x$, and

$$\pi(x) - x^b \le \pi(x) - \pi(x^b) = \sum_{x^b < p \le x} 1 \le \sum_{x^b < p \le x} \frac{\log p}{\log x^b}$$

$$= \frac{1}{b \log x} \sum_{x^b < p \le x} \log p \le \frac{1}{b \log x} \sum_{p^a \le x} \log p$$

$$= \frac{1}{b \log x} \psi(x).$$

Therefore,

(21)
$$\pi(x) \le \frac{\psi(x)}{b \log x} + x^b$$

and

$$\frac{\psi(x)}{x} \le \frac{\pi(x) \log x}{x} \le \frac{\psi(x)}{bx} + \frac{\log x}{x^{1-b}}.$$

Now take limits as $x \to \infty$; by Corollary 44.2,

$$1 = \lim_{x \to \infty} \frac{\psi(x)}{x} \le \lim \frac{\pi(x)}{x/\log x} \le \lim \frac{\psi(x)}{bx} + \lim \frac{\log x}{x^{1-b}} = 1 + 0.$$

Hence, $\pi(x) \sim x/\log x$.

Theorem 45.2. *The following are equivalent*:

(22)
$$\pi(x) \sim \frac{x}{\log x}$$

(23)
$$\psi(x) \sim x$$

(24)
$$\theta(x) \sim x.$$

Proof. We have already proved $(23) \Rightarrow (22)$. The reverse implication is proved easily. We get

$$\frac{b\pi(x) \log x}{x} - \frac{b \log x}{x^{1-b}} \le \frac{\psi(x)}{x} \le \frac{\pi(x) \log x}{x}$$

where the lower bound is obtained from (21). Now if we assume (22) and notice that $b \to 1$ as $x \to \infty$, upon taking limits, (23) follows.

The proof that (24) $\Leftrightarrow$ (22) is almost identical with the proofs already given. Each of the statements in the proofs of Theorems 45.1 and 45.2 can also be made with $\theta(x)$ replacing $\psi(x)$, and the details will be left as an exercise.

Our next theorem, which is also equivalent to the Prime Number Theorem, gives an asymptotic formula for the nth prime.

Theorem 45.3. *If p_n denotes the nth prime, then $p_n \sim n \log n$, and the Prime Number Theorem may be deduced from this relation.*

Proof. First, assume the Prime Number Theorem. Then

$$\frac{p_n}{\log p_n} = \pi(p_n) + o(\pi(p_n)) = n + o(n)$$

and

$$(25) \qquad p_n = n \log p_n + o(n \log p_n) = (n \log p_n)\{1 + o(1)\}.$$

Taking logs, we get

$$\log p_n = \log n + \log \log p_n + \log(1 + o(1));$$

but here, $\log \log p_n = o(\log p_n)$, and $\log(1 + o(1)) = o(1)$ since $\lim \log(1 + o(1)) = \log(\lim\{1 + o(1)\}) = \log 1 = 0$. Hence

$$\log p_n = \log n + o(\log p_n),$$

and by Exercise 44-1, $\log n \sim \log p_n$, so $n \log n \sim n \log p_n$. From (25) and Exercise 44-1, we get $n \log p_n \sim p_n$, and by transitivity we get $n \log n \sim p_n$.

Conversely, assume that $p_n \sim n \log n$. This implies (25), and from that, as before, we can derive the relation $\log p_n \sim \log n$. For any x, we find primes p_n, p_{n+1} such that

$$p_n \le x < p_{n+1}.$$

Hence $\pi(x) = n$. Since the function $f(y) = y/\log y$ is an increasing function, we see that

$$\frac{p_n}{\log p_n} \le \frac{x}{\log x} < \frac{p_{n+1}}{\log p_{n+1}}.$$

We divide through by $n = \pi(x)$, and get

$$\frac{p_n}{n \log p_n} \le \frac{x}{\pi(x) \log x} < \frac{p_{n+1}}{n \log p_{n+1}}.$$

Now take limits as $n \to \infty$ (then also $x \to \infty$) and we have

$$\frac{p_n}{n \log p_n} \sim \frac{n \log n}{n \log n} = 1 \quad \text{and} \quad \frac{p_{n+1}}{n \log p_{n+1}} \sim \frac{(n+1) \log(n+1)}{n \log(n+1)} \sim 1.$$

Hence,

$$\frac{x}{\pi(x) \log x} \to 1 \quad \text{as } x \to \infty,$$

and this is the Prime Number Theorem.

For every $n \in \mathcal{Z}^+$, there exists a prime p such that $n \le p \le 2n$. This fact is known as *Bertrand's conjecture*. We will give only an indication of the proof here, but instead we will prove a related result in Theorem 45.4. Since $\theta(x) \sim x$, it is obvious that there are constants C_6 and C_7 such that

$$(26) \qquad\qquad C_6 x < \theta(x) < C_7 x$$

for x sufficiently large. Since $\theta(x)$ is a step function with a (positive) jump only at the primes, it suffices to prove that $\theta(2n) - \theta(n) > 0$ for all n. But for n sufficiently large, from (26), $\theta(2n) - \theta(n) > (2C_6 - C_7)n$, and by estimating the values of the constants, one proves that $2C_6 - C_7 > 0$; then by direct verification, the result is extended to those n not included in the argument.

Our next theorem is a generalization of Bertrand's conjecture, since the latter is obtained as a special case of Theorem 45.4 by taking $\varepsilon = 1$ and $N = 1$. However, Theorem 45.4 gives no estimate of the value of x_0, while in the Bertrand conjecture $x_0 = 1$.

Theorem 45.4. *Let ε be any positive real number and N be any positive integer. Then there exists x_0 (depending on ε and on N) such that if $x \ge x_0$, the interval $[x, (1 + \varepsilon)x]$ contains at least N primes.*

Proof. We must show that for x sufficiently large, $\pi((1 + \varepsilon)x) - \pi(x) \ge N$. But

$$\pi((1 + \varepsilon)x) - \pi(x) = \frac{(1 + \varepsilon)x}{\log(1 + \varepsilon)x} - \frac{x}{\log x} + o\!\left(\frac{(1 + \varepsilon)x}{\log(1 + \varepsilon)x}\right) + o\!\left(\frac{x}{\log x}\right).$$

(27)

We will show that both little-o terms are $o(\varepsilon x/\log(1 + \varepsilon)x)$ and the dominant term in (27) is of the same form.

First, if

$$f(x) = o\!\left(\frac{(1 + \varepsilon)x}{\log(1 + \varepsilon)x}\right)$$

then

$$0 = \lim_{x \to \infty} \frac{f(x) \log(1 + \varepsilon)x}{(1 + \varepsilon)x} = \frac{1}{1 + \varepsilon} \lim_{x \to \infty} \frac{f(x) \log(1 + \varepsilon)x}{x}$$

so that the limit is zero. But then

$$0 = \frac{1}{\varepsilon} \lim_{x \to \infty} \frac{f(x) \log(1 + \varepsilon)x}{x} = \lim_{x \to \infty} \frac{f(x)}{\varepsilon x/\log(1 + \varepsilon)x},$$

so

$$f(x) = o\left(\frac{\varepsilon x}{\log(1 + \varepsilon)x}\right).$$

Also

$$0 \le \frac{|o(x/\log x)|}{\varepsilon x/\log(1 + \varepsilon)x} = \left\{\frac{1}{\varepsilon} + \frac{\log(1 + \varepsilon)}{\varepsilon \log x}\right\}\frac{|o(x/\log x)|}{x/\log x} \to 0,$$

so

$$o(x/\log x) = o(\varepsilon x/\log(1 + \varepsilon)x).$$

Finally,

$$\frac{(1 + \varepsilon)x}{\log(1 + \varepsilon)x} - \frac{x}{\log x} = \frac{\varepsilon x}{\log(1 + \varepsilon)x} - \frac{\varepsilon x}{\log(1 + \varepsilon)x}\frac{\log(1 + \varepsilon)}{\varepsilon \log x}$$

$$= \frac{\varepsilon x}{\log(1 + \varepsilon)x} + o\left(\frac{\varepsilon x}{\log(1 + \varepsilon)x}\right).$$

Making substitutions into (27), we conclude

$$\pi((1 + \varepsilon)x) - \pi(x) \sim \frac{\varepsilon x}{\log(1 + \varepsilon)x}.$$

But the function on the right $\to \infty$, so for x sufficiently large, $\pi((1 + \varepsilon)x) - \pi(x) \ge N$.

Our last application of the Prime Number Theorem is concerned with the function Li(x), called the "logarithmic integral" function, and defined for all $x > 1$ by

$$\text{Li}(x) = \int_2^x \frac{dt}{\log t}.$$

Theorem 45.5.

$$\text{Li}(x) \sim \pi(x)$$

Proof. We will show that Li$(x) \sim x/\log x$, and then this theorem follows by a trivial application of Theorem 45.1. Our proof uses the following fact about the Riemann integral of a continuous function: if $f(t)$ is continuous for all t such that $a \le t \le b$, and if $f(t)$ is positive for $a \le t \le b$, then

$$(b - a) \min_{a \le t \le b} f(t) \le \int_a^b f(t)\, dt \le (b - a) \max_{a \le t \le b} f(t).$$

It follows from this that such an integral is always non-negative, since $0 \le (b - a) \min f(t)$.

Evidently $1/\log t$ is positive and decreasing for all $t > 1$. Hence

$$\frac{x-2}{\log x} \leq \mathrm{Li}(x) \leq \frac{x-2}{\log 2}.$$

For our purpose [to show that $\mathrm{Li}(x) \sim x/\log x$], this upper bound is not satisfactory, so we must refine our method.

For each fixed $n \in \mathscr{Z}^+$,

$$\frac{x}{2^n \log x} \left\{ \frac{1}{1 - \dfrac{(n-1)\log 2}{\log x}} \right\} = \frac{x/2^{n-1} - x/2^n}{\log(x/2^{n-1})}$$

$$\leq \int_{x/2^n}^{x/2^{n-1}} \frac{dt}{\log t}$$

$$\leq \frac{x/2^{n-1} - x/2^n}{\log(x/2^n)}$$

$$= \frac{x}{2^n \log x} \left\{ \frac{1}{1 - \dfrac{n \log 2}{\log x}} \right\}.$$

Since the integral appearing above is $\mathrm{Li}(x/2^{n-1}) - \mathrm{Li}(x/2^n)$, we get

$$\frac{1}{1 - \dfrac{(n-1)\log 2}{\log x}} \leq \frac{\mathrm{Li}\left(\dfrac{x}{2^{n-1}}\right) - \mathrm{Li}\left(\dfrac{x}{2^n}\right)}{\dfrac{x}{2^n \log x}} \leq \frac{1}{1 - \dfrac{n \log 2}{\log x}}.$$

Taking limits as $x \to \infty$, we prove that

$$\mathrm{Li}\left(\frac{x}{2^{n-1}}\right) - \mathrm{Li}\left(\frac{x}{2^n}\right) \sim \frac{x}{2^n \log x}.$$

By induction on n we will show that we can add corresponding members of the above asymptotic relation and preserve the relation.

$$\frac{\mathrm{Li}(x) - \mathrm{Li}(x/2^2)}{\dfrac{3}{4} \dfrac{x}{\log x}} = \frac{\mathrm{Li}(x) - \mathrm{Li}(x/2)}{\dfrac{3}{4} \dfrac{x}{\log x}} + \frac{\mathrm{Li}(x/2) - \mathrm{Li}(x/2^2)}{\dfrac{3}{4} \dfrac{x}{\log x}}$$

$$= \frac{1}{3} \frac{\mathrm{Li}(x) - \mathrm{Li}(x/2)}{\dfrac{1}{2} \dfrac{x}{\log x}} + \frac{1}{3} \frac{\mathrm{Li}(x/2) - \mathrm{Li}(x/4)}{\dfrac{1}{4} \dfrac{x}{\log x}}$$

$$\to \frac{2}{3} \cdot 1 + \frac{1}{3} \cdot 1 = 1.$$

Therefore, $\text{Li}(x) - \text{Li}(x/4) \sim 3x/4 \log x$. Assume that for $k \geq 2$,

(28) $$\text{Li}(x) - \text{Li}\left(\frac{x}{2^k}\right) \sim \frac{2^k - 1}{2^k} \frac{x}{\log x}.$$

Then

$$\frac{\text{Li}(x) - \text{Li}(x/2^{k+1})}{\dfrac{2^{k+1} - 1}{2^{k+1}} \dfrac{x}{\log x}} = \frac{2(2^k - 1)}{2^{k+1} - 1} \frac{\text{Li}(x) - \text{Li}(x/2^k)}{\dfrac{2^k - 1}{2^k} \dfrac{x}{\log x}}$$

$$+ \frac{1}{2^{k+1} - 1} \frac{\text{Li}(x/2^k) - \text{Li}(x/2^{k+1})}{\dfrac{1}{2^{k+1}} \dfrac{x}{\log x}}$$

$$\to \frac{2(2^k - 1)}{2^{k+1} - 1} + \frac{1}{2^{k+1} - 1} = 1.$$

Therefore,

$$\text{Li}(x) - \text{Li}(x/2^{k+1}) \sim \frac{2^{k+1} - 1}{2^{k+1}} \frac{x}{\log x},$$

the induction is complete, and (28) holds for all $k \in \mathscr{Z}^+$.

Let $a + 1 = [\log x/\log 2]$, so $2^{a+1} \leq x < 2^{a+2}$, and $2 \leq x/2^a < 4$. By (28), with $k = a$,

$$\text{Li}(x) - \text{Li}\left(\frac{x}{2^a}\right) \sim \frac{2^a - 1}{2^a} \frac{x}{\log x} \sim \frac{x}{\log x}$$

since as $x \to \infty$, then $a \to \infty$, and $(2^a - 1)/2^a \to 1$. But

$$\text{Li}\left(\frac{x}{2^a}\right) = \int_2^{x/2^a} \frac{dt}{\log t} \leq \int_2^4 \frac{dt}{\log t} = 0(1)$$

so that

$$1 = \lim_{x \to \infty} \left(\frac{\text{Li}(x)}{x/\log x} - \frac{\text{Li}(x/2^a)}{x/\log x}\right)$$

$$= \lim \frac{\text{Li}(x)}{x/\log x} + \lim \frac{0(1)}{x/\log x} = \lim \frac{\text{Li}(x)}{x/\log x} + 0.$$

This proves $\text{Li}(x) \sim x/\log x$, so $\text{Li}(x) \sim \pi(x)$.

EXERCISES

45-1. Complete the proof of Theorem 45.2 by showing that $(24) \Leftrightarrow (23)$.

45-2. If p_n denotes the nth prime, prove $p_n \sim p_{n+1}$.

45-3. Prove that Theorem 45.5 is equivalent to the Prime Number Theorem.

Chapter 9

GEOMETRY OF NUMBERS

FOR THE MATERIAL in this chapter we assume that the reader has some background in linear algebra. In order to minimize the requirement for a knowledge of linear algebra, we will only study geometry of numbers in the plane; furthermore, our intuition can aid our understanding since we are confining our attention to the 2-dimensional case. Most of the results of this chapter can readily be extended to n-dimensional Euclidean spaces.

The geometry of numbers was first studied by Minkowski. We shall prove some of the fundamental results of Minkowski and apply the results to obtain an upper bound on the least positive value attained by a positive definite binary quadratic form. We will take a second look at Farey sequences from a geometric point of view, and we end our study by using geometric methods to prove that the infinite continued fraction expansion of α is periodic if and only if α is a quadratic irrational.

46. Preliminaries

Let E_2 denote the Cartesian plane. The elements of E_2 will be called *points* or *vectors* and will be denoted by the letters $c, d, p, q, \ldots$. Throughout the chapter the vectors a, b, and o will always mean

$$a = (1,0),$$

$$b = (0,1),$$

$$o = (0,0).$$

If $p, q \in E_2$, say $p = (p_1, p_2)$ and $q = (q_1, q_2)$, then $p = q$ if and only if $p_i = q_i$ $(i = 1,2)$, and the vector $p + q$, called the *vector sum* of p and q, is the vector

$$p + q = (p_1 + q_1, p_2 + q_2);$$

if x is any real number, then $x\mathbf{p}$ is the vector

$$x\mathbf{p} = (xp_1, xp_2).$$

The vectors $\mathbf{p}$ and $\mathbf{q}$ are said to be *linearly independent* provided that $x\mathbf{p} + y\mathbf{q} = \mathbf{o}$ implies $x = y = 0$. Hence $\mathbf{p}$ and $\mathbf{q}$ are *linearly dependent* if and only if there exists $(x,y) \neq \mathbf{o}$ such that $x\mathbf{p} + y\mathbf{q} = \mathbf{o}$.

If $\mathbf{p}$ and $\mathbf{q}$ are fixed points in $\mathbf{E}_2$, we define the set $\Lambda(\mathbf{p},\mathbf{q}) \subset \mathbf{E}_2$ by

$$\Lambda(\mathbf{p},\mathbf{q}) = \{m\mathbf{p} + n\mathbf{q} : m, n \in \mathscr{Z}\}.$$

Thus, $\Lambda(\mathbf{p},\mathbf{q})$ is the set of certain linear combinations of $\mathbf{p}$ and $\mathbf{q}$, and is said to be *generated by* $\mathbf{p}$ and $\mathbf{q}$. The set $\Lambda(\mathbf{p},\mathbf{q})$ will be called an *integral lattice* if and only if $\mathbf{p}$ and $\mathbf{q}$ are linearly independent, and the points $m\mathbf{p} + n\mathbf{q}$ in an integral lattice are called *lattice points*.

Example. With $\mathbf{p} = \mathbf{a}$ and $\mathbf{q} = \mathbf{b}$ we have

$$\Lambda(\mathbf{a},\mathbf{b}) = \{m\mathbf{a} + n\mathbf{b} : m, n \in \mathscr{Z}\}$$

$$= \{(m,n) : m, n \in \mathscr{Z}\}.$$

This integral lattice is just the set of "lattice points" in the sense of Section 17, i.e., the set of all points with integer coordinates. Clearly this Λ is an integral lattice because if $m\mathbf{a} + n\mathbf{b} = (m,n) = \mathbf{o}$, then $m = n = 0$, so $\mathbf{a}$ and $\mathbf{b}$ are linearly independent.

Let $\Lambda_1 = \Lambda(\mathbf{a},\mathbf{b})$ and $\Lambda_2 = \Lambda((1,1),(2,3))$. We notice that $\Lambda_1 = \Lambda_2$: it is obvious that $\Lambda_2 \subset \Lambda_1$ because if $m(1,1) + n(2,3) = (m + 2n, m + 3n) \in \Lambda_2$, then the coordinates $m + 2n$ and $m + 3n$ are integers, so the point is in Λ_1; conversely, if $(m,n) \in \Lambda_1$, then we find $u, v \in \mathscr{Z}$ given by $u = 3m - 2n$, $v = n - m$, and then $(m,n) = u(1,1) + v(2,3) \in \Lambda_2$. Here it was possible to find integers u, v as required because there is a solution in integers x, y of the system of equations

(1)
$$\begin{cases} x + 2y = m \\ x + 3y = n. \end{cases}$$

Evidently it was fortunate that the determinant of coefficients in (1), namely

$$\begin{vmatrix} 1 & 2 \\ 1 & 3 \end{vmatrix},$$

has the value 1, and we notice that the determinant has for its columns the coordinates of the vectors which generate Λ_2. These observations shall guide us in introducing certain definitions.

Since the point-pairs $\mathbf{a},\mathbf{b}$ and $(1,1)$, $(2,3)$ generate the same integral lattice, it is natural to regard the point-pairs as related in some sense. We will say the point-pairs **p,q** and **m,n** are *equivalent*, and we will write **p,q** $\simeq$ **m,n** if and only if $\Lambda(\mathbf{p,q}) = \Lambda(\mathbf{m,n})$. It is clear that " $\simeq$ " is an equivalence relation on $\mathbf{E}_2 \times \mathbf{E}_2$, and hence that it partitions the set into equivalence classes. We shall obtain a characterization of these classes.

An integral lattice $\Lambda(\mathbf{p,q})$ may be viewed geometrically. Consider the vectors **p** and **q** in the plane, that is, the two directed line segments from **o** to **p** and from **o** to **q** (see Figure 6). Let L_0 and L_0' be the lines containing the

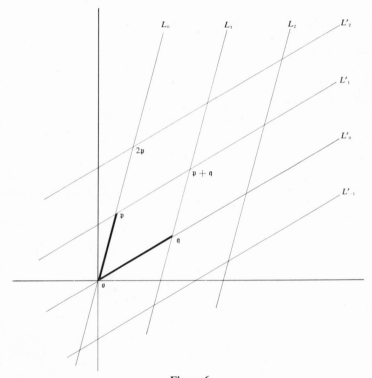

Figure 6

vectors **p** and **q**, respectively. For $j \in \mathscr{Z}$, let L_j be the line parallel to L_0 and through the point $j\mathbf{q}$; let L_j' be the line parallel to L_0' and through $j\mathbf{p}$. Now the plane is covered by two systems of parallel lines; the collection of points of intersection of these parallel systems is the set of points of $\Lambda(\mathbf{p,q})$, or

$$\{L_j \cap L_k' : j, k \in \mathscr{Z}\} = \Lambda(\mathbf{p,q}).$$

If M is any matrix, then det M will denote the *determinant* of M, tr M will be the *transpose* of M, and M^{-1} will denote the *inverse* of M if the inverse exists. We call M an *integral matrix* if the entries of M are integers. If $\mathbf{p},\mathbf{q} \in E_2$, we let $(\mathbf{p}\ \mathbf{q})$ denote the matrix whose first column consists of the coordinates of $\mathbf{p}$, and whose second column is $\mathbf{q}$. For example, if $\mathbf{p} = (-\frac{1}{2}, \frac{3}{5})$ and $\mathbf{q} = (2, \frac{5}{4})$, then

$$(\mathbf{p}\ \mathbf{q}) = \begin{pmatrix} -\frac{1}{2} & 2 \\ \frac{3}{5} & \frac{5}{4} \end{pmatrix}$$

and $\det (\mathbf{p}\ \mathbf{q}) = -\frac{73}{40}$.

We can now prove the following theorem.

Theorem 46.1. *Suppose* $\mathbf{p},\mathbf{q},\mathbf{m},\mathbf{n} \in E_2$. *Then* $\mathbf{p},\mathbf{q} \simeq \mathbf{m},\mathbf{n}$ *if and only if there exists an integral matrix* U *with* $\det U = \pm 1$ *such that* $(\mathbf{p}\ \mathbf{q})U = (\mathbf{m}\ \mathbf{n})$.

Proof. Suppose first that $\mathbf{p},\mathbf{q} \simeq \mathbf{m},\mathbf{n}$. Then $\Lambda(\mathbf{p},\mathbf{q}) = \Lambda(\mathbf{m},\mathbf{n})$, and in particular $\mathbf{m} \in \Lambda(\mathbf{p},\mathbf{q})$ and $\mathbf{n} \in \Lambda(\mathbf{p},\mathbf{q})$. Therefore there are integers a,b,c,d such that $\mathbf{m} = a\mathbf{p} + b\mathbf{q}$ and $\mathbf{n} = c\mathbf{p} + d\mathbf{q}$. Thus, $(\mathbf{p}\ \mathbf{q})U = (\mathbf{m}\ \mathbf{n})$, with

$$U = \begin{pmatrix} a & c \\ b & d \end{pmatrix},$$

an integral matrix. In a similar manner, since $\mathbf{p},\mathbf{q} \in \Lambda(\mathbf{m},\mathbf{n})$, there exists an integral matrix V such that $(\mathbf{p}\ \mathbf{q}) = (\mathbf{m}\ \mathbf{n})V$.

Since $\mathbf{p}, \mathbf{q}$ are linearly independent, $\det (\mathbf{p}\ \mathbf{q}) \neq 0$ and $(\mathbf{p}\ \mathbf{q})^{-1}$ exists. Now we have

$$UV = (\mathbf{p}\ \mathbf{q})^{-1}[(\mathbf{p}\ \mathbf{q})U]V$$
$$= (\mathbf{p}\ \mathbf{q})^{-1}(\mathbf{m}\ \mathbf{n})V$$
$$= (\mathbf{p}\ \mathbf{q})^{-1}(\mathbf{p}\ \mathbf{q}) = I$$

where I is the 2×2 identity matrix. We know $\det U \in \mathscr{Z}$ and $\det V \in \mathscr{Z}$ since U and V are integral. But $1 = \det I = \det UV = (\det U)(\det V)$, so $\det U = \pm 1$ as was to be proved. Evidently $V = U^{-1}$ is also integral.

Conversely, assume there is an integral matrix U with determinant ± 1 such that $(\mathbf{p}\ \mathbf{q})U = (\mathbf{m}\ \mathbf{n})$. Suppose $s_1\mathbf{p} + s_2\mathbf{q} \in \Lambda(\mathbf{p},\mathbf{q})$; if we let $\mathbf{s} = (s_1,s_2)$, we are supposing that $(\mathbf{s})$ is an integral matrix and $(\mathbf{p}\ \mathbf{q})(\mathbf{s}) \in \Lambda(\mathbf{p},\mathbf{q})$. Clearly, $U^{-1}(\mathbf{s})$ is an integral matrix, say $U^{-1}(\mathbf{s}) = (\mathbf{\eth})$. But

(2) $$(\mathbf{p}\ \mathbf{q})(\mathbf{s}) = (\mathbf{p}\ \mathbf{q})U(\mathbf{\eth}) = (\mathbf{m}\ \mathbf{n})(\mathbf{\eth}),$$

so the point $(\mathbf{p}\ \mathbf{q})(\mathbf{s}) \in \Lambda(\mathbf{m},\mathbf{n})$, and $\Lambda(\mathbf{p},\mathbf{q}) \subset \Lambda(\mathbf{m},\mathbf{n})$. The reverse inclusion is easily proved from the matrix equation (2), and it follows that $\Lambda(\mathbf{m},\mathbf{n}) = \Lambda(\mathbf{p},\mathbf{q})$.

EXERCISES

46-1.   Suppose U is an integral matrix. Find necessary and sufficient conditions so that U^{-1} exists and is integral.

46-2.   Suppose $\mathbf{p},\mathbf{q}$ are linearly independent. Prove $\mathbf{p},\mathbf{q} \simeq \mathbf{q},\mathbf{p}$.

46-3.* If $\mathbf{p},\mathbf{q} \simeq \mathbf{p},\mathbf{m}$, prove that there is an integer a such that $\mathbf{q} \pm \mathbf{m} = a\mathbf{p}$.

46-4.* In $\Lambda(\mathbf{p},\mathbf{q})$, consider the parallelogram based on $\mathbf{p}$ and $\mathbf{q}$ (that is, the parallelogram with sides that are the line segments from $\mathbf{o}$ to $\mathbf{p}$, $\mathbf{p}$ to $\mathbf{p} + \mathbf{q}$, $\mathbf{p} + \mathbf{q}$ to $\mathbf{q}$, and $\mathbf{q}$ to $\mathbf{o}$). Prove this parallelogram has area $|\det (\mathbf{p}\ \mathbf{q})|$.

46-5.   Let $\Lambda = \Lambda(\mathbf{p},\mathbf{q})$. Suppose $\mathbf{m},\mathbf{n} \in \Lambda$ have the property that $|\det (\mathbf{m}\ \mathbf{n})|$ is equal to the area of the parallelogram based on $\mathbf{p}$ and $\mathbf{q}$ (see Exercise 46-4). Prove $\mathbf{m},\mathbf{n} \simeq \mathbf{p},\mathbf{q}$.

46-6.* Suppose $\mathbf{m},\mathbf{n} \in \Lambda(\mathbf{p},\mathbf{q})$. Prove that a necessary and sufficient condition for $\mathbf{m},\mathbf{n} \simeq \mathbf{p},\mathbf{q}$ is that the triangle with vertices $\mathbf{o},\mathbf{m},\mathbf{n}$ contains no point of Λ except the vertices.

46-7.   Prove there are lines which contain no points of an integral lattice Λ, and lines which contain exactly one point of Λ, but if a line contains at least two points of Λ then it contains infinitely many.

47. Convex Symmetric Distance Functions

If $\Lambda = \Lambda(\mathbf{p},\mathbf{q})$ is an integral lattice, we define $\Delta(\Lambda)$ to be $|\det (\mathbf{p}\ \mathbf{q})|$. It was shown in Exercise 46-4 that $\Delta(\Lambda)$ is the area of the parallelogram based on $\mathbf{p}$ and $\mathbf{q}$. We denote the *length* of a vector $\mathbf{p} = (p_1,p_2)$ by $|\mathbf{p}|$ and define $|\mathbf{p}| = (p_1^2 + p_2^2)^{1/2}$. Let us assume that $|\mathbf{p}| \leq |\mathbf{q}|$ and consider the circle C with center at the origin and radius $r/2 = (1/2)|\mathbf{p}|$; we would like to show that

(3) $$\pi r^2 < 4\Delta(\Lambda).$$

In Figure 7 we have a circle with radius $r/2$ at each of the vertices of the parallelogram, and it is clear that the sum of the shaded areas is exactly $\pi(r/2)^2$, the area of C. But the shaded areas cannot cover the parallelogram because of the way $r/2$ was chosen, so the area $\Delta(\Lambda)$ is larger than the area of C, which proves (3). Now in turn (3) tells us that the area πr^2 of a circle of radius $r = |\mathbf{p}|$ is less than $4\Delta(\Lambda)$; such a circle is also indicated in Figure 7.

The illustration, which shows how geometric observations can yield arithmetic facts, and vice versa, is a special case of a theorem of Minkowski. Before stating the theorem, we will generalize the important aspects of the above illustration. The circle C will be replaced by any set which is a closed, bounded, convex body symmetric with respect to $\mathbf{o}$ (these terms will be defined below). However, notice that if $f(\mathbf{s}) = |\mathbf{s}|$, then C can be described as the set of points $\mathbf{s}$ such that $f(\mathbf{s}) \leq r/2$; the function f will be replaced by a generalized distance function.

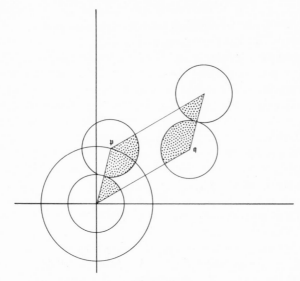

Figure 7

If $\mathscr{S}$ is a set in $\mathbf{E}_2$ and $\mathbf{p} \in \mathscr{S}$, we say $\mathbf{p}$ is an *interior point* of $\mathscr{S}$ if and only if there exists a circle C with center at $\mathbf{p}$ and such that $C \subset \mathscr{S}$. The set $\mathscr{S}$ is *open* if and only if every point of $\mathscr{S}$ is an interior point, and $\mathscr{S}$ is *closed* if and only if its complement (the set of points not in $\mathscr{S}$) is open. The set $\mathscr{S}$ is a *convex* set if and only if the line segment joining any two points of $\mathscr{S}$ lies entirely in $\mathscr{S}$; that is, if $\mathbf{p},\mathbf{q} \in \mathscr{S}$, then all points

$$(1 - t)\mathbf{p} + t\mathbf{q} \quad (0 \le t \le 1)$$

are also in $\mathscr{S}$; a convex set is called a *convex body* if it contains an interior point. The set $\mathscr{S}$ is *bounded* if there is a circle with center at $\mathbf{o}$ which contains $\mathscr{S}$, and $\mathscr{S}$ is said to be *symmetric with respect to* $\mathbf{o}$ if $\mathbf{p} \in \mathscr{S}$ implies $-\mathbf{p} \in \mathscr{S}$.

Examples. If $\mathscr{S}$ is a circle, every point of $\mathscr{S}$ not on the circumference is an interior point, and points on the circumference are not interior points. A line has no interior points. A set consisting of a single point has no interior point.

The set of points inside but not on a circle is open; the sets $\mathbf{E}_2$ and $\varnothing$ are open; the set of $\mathbf{s} \ne \mathbf{o}$ is open. The interval $(0,1)$ is not open. The set of points inside or on a circle is closed; the sets $\mathbf{E}_2$ and $\varnothing$ are closed; a set with a single point is closed; an integral lattice is closed. The interval $(0,1)$ is not closed.

The set of points inside a circle is convex; the sets $\mathbf{E}_2$ and $\varnothing$ are convex; any line is a convex set; the set of points in a triangle is convex. A set consisting of a circle and a line is not convex; a set of two points is not convex. The points inside a circle form a convex body; E_2 is a convex body, $\varnothing$ is not a convex body; a line is not a convex body.

An integral lattice is symmetric with respect to $\mathbf{o}$, and so are the sets E_2 and $\varnothing$. A circle with center at $\mathbf{a}$ is not symmetric with respect to $\mathbf{o}$.

Definition. *Suppose f is a real-valued function defined for all points $\mathbf{p} \in E_2$. We call f a* convex symmetric distance function (csdf) *if*

(4) $f(\mathbf{p}) \geq 0$ *for all $\mathbf{p} \in E_2$, and $f(\mathbf{p}) = 0$ if and only if $\mathbf{p} = \mathbf{o}$;*

(5) $f(t\mathbf{p}) = |t| f(\mathbf{p})$ *for all real t, all $\mathbf{p} \in E_2$;*

(6) $f(\mathbf{p} + \mathbf{q}) \leq f(\mathbf{p}) + f(\mathbf{q})$, *for all $\mathbf{p}, \mathbf{q} \in E_2$.*

There certainly are such functions because $f(\mathbf{p}) = |\mathbf{p}|$ has the required properties for a csdf. We will show that every csdf is a continuous function. If f is a csdf, then f is continuous at $\mathbf{o}$, because if $\mathbf{p}_1, \mathbf{p}_2, \ldots$ is a sequence of points approaching $\mathbf{o}$, say $\mathbf{p}_n = (p_{1n}, p_{2n}) = p_{1n}\mathbf{a} + p_{2n}\mathbf{b}$, then

$$f(\mathbf{p}_n) \leq |p_{1n}| f(\mathbf{a}) + |p_{2n}| f(\mathbf{b}).$$

Let $m = \max (f(\mathbf{a}), f(\mathbf{b}))$; then $0 \leq f(\mathbf{p}_n) \leq m(|p_{1n}| + |p_{2n}|)$ and as $\mathbf{p}_n \to \mathbf{o}$, $|p_{1n}| + |p_{2n}| \to 0$, so $f(\mathbf{p}_n) \to f(\mathbf{o}) = 0$, which proves f is continuous at $\mathbf{o}$. Now if $\mathbf{p}_0 \in E_2$, we prove f is continuous at $\mathbf{p}_0$. Let $\mathbf{p}_n \to \mathbf{p}_0$, so $\mathbf{p}_n - \mathbf{p}_0 \to \mathbf{o}$. In (6), we replace $\mathbf{p}$ by $\mathbf{m} - \mathbf{n}$ and $\mathbf{q}$ by $\mathbf{n}$ to see that $f(\mathbf{m} - \mathbf{n}) \geq f(\mathbf{m}) - f(\mathbf{n})$ for all $\mathbf{m}, \mathbf{n} \in E_2$; and in particular,

(7) $f(\mathbf{p}_n - \mathbf{p}_0) \geq f(\mathbf{p}_n) - f(\mathbf{p}_0).$

But from $f(\mathbf{p}_n - \mathbf{p}_0) = f(-(\mathbf{p}_0 - \mathbf{p}_n)) = |-1| f(\mathbf{p}_0 - \mathbf{p}_n) \geq f(\mathbf{p}_0) - f(\mathbf{p}_n)$, we have $f(\mathbf{p}_n) - f(\mathbf{p}_0) \geq -f(\mathbf{p}_n - \mathbf{p}_0)$; together with (7) we get

$$-f(\mathbf{p}_n - \mathbf{p}_0) \leq f(\mathbf{p}_n) - f(\mathbf{p}_0) \leq f(\mathbf{p}_n - \mathbf{p}_0).$$

Since $\mathbf{p}_n - \mathbf{p}_0 \to \mathbf{o}$ and f is continuous at $\mathbf{o}$, we have $f(\mathbf{p}_n) - f(\mathbf{p}_0) \to 0$, or $f(\mathbf{p}_n) \to f(\mathbf{p}_0)$. Therefore, f is continuous.

If $k > 0$, we consider the set of all $\mathbf{p}$ such that $f(\mathbf{p}) \leq k$. Since we are interested only in topological properties of the set, we may assume $k = 1$ since $f(\mathbf{p}) \leq k$ if and only if

$$f\left(\frac{1}{k}\mathbf{p}\right) = \frac{1}{k}f(\mathbf{p}) \leq 1.$$

Theorem 47.1. *Suppose f is a csdf and let*

$$\mathscr{S} = \{\mathbf{p} \in E_2 : f(\mathbf{p}) \leq 1\}.$$

Then $\mathscr{S}$ is a closed, bounded, convex body symmetric with respect to $\mathbf{o}$. A point $\mathbf{p}$ is an interior point of $\mathscr{S}$ if and only if $f(\mathbf{p}) < 1$.

Proof. It is immediate from (5) that $\mathscr{S}$ is symmetric, because $f(-\mathbf{p}) = |-1| f(\mathbf{p})$, so $\mathbf{p} \in \mathscr{S}$ if and only if $-\mathbf{p} \in \mathscr{S}$. The set is convex, because if $\mathbf{p},\mathbf{q} \in \mathscr{S}$ and t is real, $0 \le t \le 1$, then

$$f((1 - t)\mathbf{p} + t\mathbf{q}) \le |1 - t| f(\mathbf{p}) + |t| f(\mathbf{q}) \le |1 - t| + |t| = 1 ;$$

hence $(1 - t)\mathbf{p} + t\mathbf{q} \in \mathscr{S}$. But $\mathscr{S}$ also contains interior points, because if $f(\mathbf{p}) < 1$, by the continuity of f, points $\mathbf{q}$ sufficiently close to $\mathbf{p}$ have values $f(\mathbf{q})$ close to $f(\mathbf{p})$. More precisely, there is some $\delta > 0$ such that if $|\mathbf{q} - \mathbf{p}| < \delta$, then $|f(\mathbf{q}) - f(\mathbf{p})| < 1 - f(\mathbf{p})$, or $f(\mathbf{q}) < 1$; but the set of $\mathbf{q}$ such that $|\mathbf{q} - \mathbf{p}| < \delta$ is a circle with center at $\mathbf{p}$, and all such $\mathbf{q}$ satisfy $f(\mathbf{q}) < 1$, so $\mathbf{q} \in \mathscr{S}$. Therefore, $\mathscr{S}$ is a convex body, and we have also proved that if $f(\mathbf{p}) < 1$, then $\mathbf{p}$ is an interior point.

There are points $\mathbf{p}$ which are not in $\mathscr{S}$; for example, since $f(\mathbf{a}) > 0$ and $f(t\mathbf{a}) = |t| f(\mathbf{a})$, we see that the point $t\mathbf{a} \notin \mathscr{S}$ if $t > 1/f(\mathbf{a})$. If $\mathbf{p} \notin \mathscr{S}$, then $f(\mathbf{p}) > 1$ and, again by the continuity of f, if $\mathbf{q}$ is close to $\mathbf{p}$, then $f(\mathbf{q})$ is close to $f(\mathbf{p})$, so that $\mathbf{q} \notin \mathscr{S}$ for all $\mathbf{q}$ sufficiently close to $\mathbf{p}$. Therefore, if $\mathbf{p} \notin \mathscr{S}$, then $\mathbf{p}$ is an interior point of the complement of $\mathscr{S}$, so $\mathscr{S}$ is closed.

To prove boundedness, let Γ denote the unit circle in $\mathbf{E}_2$, that is, $\Gamma = \{\mathbf{p} : \mathbf{p} \in \mathbf{E}_2, |\mathbf{p}| = 1\}$. Clearly Γ is closed and bounded, so f (a continuous function on a closed and bounded set) attains a minimum, say m, on Γ; hence $m \le f(\mathbf{p})$ for all $\mathbf{p} \in \Gamma$. If $\mathbf{p} \ne \mathbf{o}$, then

$$\frac{\mathbf{p}}{|\mathbf{p}|} \in \Gamma,$$

so by (5) we have

$$m \le \frac{1}{|\mathbf{p}|} f(\mathbf{p}) \quad \text{for all } \mathbf{p} \in \mathbf{E}_2,$$

Now $m \ne 0$ because for some $\mathbf{m} \in \Gamma$, we have $f(\mathbf{m}) = m$, and since $\mathbf{m} \in \Gamma$ implies $\mathbf{m} \ne \mathbf{o}$, it follows that $m = f(\mathbf{m}) \ne 0$. Therefore, $|\mathbf{p}| \le f(\mathbf{p})/m$ for all $\mathbf{p} \in \mathbf{E}_2$; hence, if $\mathbf{p} \in \mathscr{S}$, $|\mathbf{p}| \le 1/m$ and $\mathscr{S}$ is bounded.

It remains only to show that if $\mathbf{p}$ is an interior point of $\mathscr{S}$, then $f(\mathbf{p}) < 1$. We prove this by showing that if $f(\mathbf{p}) = 1$, then $\mathbf{p}$ is not an interior point of $\mathscr{S}$, because for every $\varepsilon > 0$, $f((1 + \varepsilon)\mathbf{p}) = (1 + \varepsilon) > 1$, so the points $(1 + \varepsilon)\mathbf{p} \notin \mathscr{S}$. But every neighborhood of $\mathbf{p}$ contains the point $(1 + \varepsilon)\mathbf{p}$ for ε sufficiently small, so $\mathbf{p}$ is not an interior point of $\mathscr{S}$. This completes the proof of Theorem 47.1.

We will show next that if $\mathscr{K}$ is any closed, bounded, symmetric, convex body, then we can associate with $\mathscr{K}$ a csdf. The details of the proof are a bit tedious, but the essential idea is simple enough. If $\mathbf{p}$ is any point in $\mathbf{E}_2$, then the set of all points $t\mathbf{p}$, $0 \leq t$, is a half line from $\mathbf{o}$ and through $\mathbf{p}$. As we move along this line from $\mathbf{o}$ we must come to a point $t_0\mathbf{p}$ which is the last point in $\mathscr{K}$. We will then take $f(t_0\mathbf{p}) = 1$ and $f(r\mathbf{p}) = r/t_0$, so that $f(r\mathbf{p})$ is ≤ 1 or > 1 according as $r \leq t_0$ or $r > t_0$; thus, $f(r\mathbf{p}) \leq 1$ if $r\mathbf{p}$ is between $\mathbf{o}$ and $t_0\mathbf{p}$, otherwise $f(r\mathbf{p}) > 1$.

We first prove the following.

Lemma 47.1. *If $\mathscr{K}$ is a convex body symmetric with respect to $\mathbf{o}$, then $\mathbf{o}$ is an interior point of $\mathscr{K}$.*

Proof. Since $\mathscr{K}$ is a convex body, it contains an interior point $\mathbf{p}$. If $\mathbf{p} = \mathbf{o}$, we are through, so assume $\mathbf{p} \neq \mathbf{o}$. Then $-\mathbf{p} \in \mathscr{K}$, and by convexity $(1 - \frac{1}{2})\mathbf{p} + \frac{1}{2}(-\mathbf{p}) = \mathbf{o} \in \mathscr{K}$. There is a circular disc Γ_1 with center at $\mathbf{p}$ which is in $\mathscr{K}$, say $\Gamma_1 = \{\mathbf{q} : |\mathbf{q} - \mathbf{p}| < \delta\}$. If $\mathbf{q} \in \Gamma_1 \subset \mathscr{K}$, then $-\mathbf{q} \in \mathscr{K}$, so the disc $\Gamma_2 = \{-\mathbf{q} : \mathbf{q} \in \Gamma_1\} = \{\mathbf{q} : |\mathbf{q} + \mathbf{p}| < \delta\} \subset \mathscr{K}$. (See Figure 8.)

The point $\mathbf{s}_1 = \mathbf{p} - (\delta/2)\mathbf{b} \in \Gamma_1$ since $|\mathbf{p} - \mathbf{s}_1| = \delta/2$. Similarly, the point $\mathbf{s}_2 = \mathbf{p} + (\delta/2)\mathbf{b} \in \Gamma_1$; hence $-\mathbf{s}_1, -\mathbf{s}_2 \in \Gamma_2$, so the line segments L_1 and L_2 are in $\mathscr{K}$, where L_1 is from $\mathbf{s}_1$ to $-\mathbf{s}_2$ and L_2 is from $\mathbf{s}_2$ to $-\mathbf{s}_1$. Now consider the region $\mathscr{T}$ (shaded in Figure 8) bounded by the discs Γ_1, Γ_2 and the lines L_1, L_2; we show $\mathscr{T} \subset \mathscr{K}$.

If $\mathbf{m} \in \mathscr{T}$, consider a line through $\mathbf{m}$ and parallel to L_1; such a line intersects points in Γ_1 and points in Γ_2, and the segment of such a line from Γ_1 to Γ_2 is

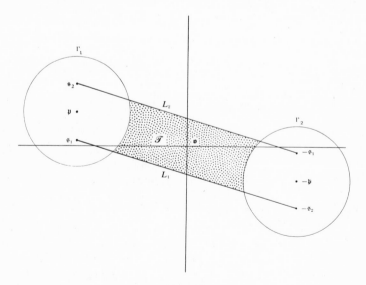

Figure 8

in $\mathcal{K}$ by convexity. Hence $\mathbf{m} \in \mathcal{K}$, so $\mathcal{T} \subset \mathcal{K}$. Evidently we can now find a circle with center at $\mathbf{o}$ and which is in $\mathcal{T}$, hence in $\mathcal{K}$. Therefore, $\mathbf{o}$ is an interior point of $\mathcal{K}$.

Theorem 47.2. *Suppose $\mathcal{K}$ is a closed and bounded convex body symmetric with respect to $\mathbf{o}$. Define a function f on $\mathbf{E}_2$ by*

$$f(\mathbf{p}) = \operatorname*{glb}_{\substack{t > 0 \\ t^{-1}\mathbf{p} \in \mathcal{K}}} t;$$

then f is a csdf. Furthermore, if $\mathcal{S}$ is the set associated with f as in Theorem 47.1, then $\mathcal{S} = \mathcal{K}$.

Proof. We know from the lemma that there is a circle Γ_3, center at $\mathbf{o}$, $\Gamma_3 \subset \mathcal{K}$, and since $\mathcal{K}$ is bounded there is a circle Γ_4 such that $\mathcal{K} \subset \Gamma_4$. Suppose the radius of Γ_i is r_i ($i = 3, 4$).

Since $t\mathbf{o} \in \mathcal{K}$ for all $t > 0$, $f(\mathbf{o}) = \operatorname{glb} t = 0$. If $\mathbf{p} \neq \mathbf{o}$, and if $0 < |t^{-1}\mathbf{p}| < r_3$, then $t^{-1}\mathbf{p} \in \Gamma_3 \subset \mathcal{K}$; thus, $t^{-1}\mathbf{p} \in \mathcal{K}$ if $t > |\mathbf{p}|/r_3$, so $f(\mathbf{p}) \leq |\mathbf{p}|/r_3$ is real. Similarly if $t < |\mathbf{p}|/r_4$, then $t^{-1}\mathbf{p} \notin \Gamma_4$, hence $\notin \mathcal{K}$, so $f(\mathbf{p}) \geq |\mathbf{p}|/r_4 > 0$. We have proved that $f(\mathbf{p})$ is real-valued, $f(\mathbf{p}) \geq 0$ for all $\mathbf{p}$, and $f(\mathbf{p}) = 0$ if and only if $\mathbf{p} = \mathbf{o}$.

To prove that $f(k\mathbf{p}) = |k| f(\mathbf{p})$, we use the fact that if $\mathscr{X}, \mathscr{Y}$ are sets of reals and $\mathscr{X} \subset \mathscr{Y}$, then $\operatorname{glb} \mathscr{X} \geq \operatorname{glb} \mathscr{Y}$. Let $\mathscr{A} = \{t : t^{-1}(k\mathbf{p}) \in \mathcal{K}\}$ and $\mathscr{B} = \{r : r^{-1}\mathbf{p} \in \mathcal{K}\}$. If $t \in \mathscr{A}$, then $t|k|^{-1} \in \mathscr{B}$; therefore,

$$|k|^{-1} f(k\mathbf{p}) = |k|^{-1} \operatorname*{glb}_{t \in \mathscr{A}} t = \operatorname*{glb}_{t \in \mathscr{A}} t|k|^{-1} \geq \operatorname*{glb}_{t|k|^{-1} \in \mathscr{B}} t|k|^{-1} = f(\mathbf{p}).$$

Also, if $r \in \mathscr{B}$, then $r|k| \in \mathscr{A}$, so

$$|k| f(\mathbf{p}) = |k| \operatorname*{glb}_{r \in \mathscr{B}} r \geq \operatorname*{glb}_{r|k| \in \mathscr{A}} r|k| = f(k\mathbf{p}).$$

We now prove that $f(\mathbf{p}) \leq 1$ if and only if $\mathbf{p} \in \mathcal{K}$. If $f(\mathbf{p}) < 1$ and $\mathbf{p} \neq \mathbf{o}$, there is some t such that $0 < f(\mathbf{p}) \leq t < 1$ for which $t^{-1}\mathbf{p} \in \mathcal{K}$. Since $\mathbf{o} \in \mathcal{K}$, convexity implies $(1 - t)\mathbf{o} + t(t^{-1}\mathbf{p}) = \mathbf{p} \in \mathcal{K}$. If $f(\mathbf{p}) = 1$, then $\mathbf{p}$ is the limit of a sequence of points $\mathbf{q}$ with $f(\mathbf{q}) < 1$; thus, $\mathbf{p}$ is the limit of a sequence of points of $\mathcal{K}$, and $\mathcal{K}$ is closed, so $\mathbf{p} \in \mathcal{K}$. Conversely, if $f(\mathbf{p}) > 1$, then from the definition of f we have $\mathbf{p} = 1^{-1}\mathbf{p} \notin \mathcal{K}$.

Finally, we must prove f satisfies $f(\mathbf{p} + \mathbf{q}) \leq f(\mathbf{p}) + f(\mathbf{q})$ for all $\mathbf{p}, \mathbf{q}$; since the proof is otherwise trivial, assume $\mathbf{p} \neq \mathbf{o}$, $\mathbf{q} \neq \mathbf{o}$, $\mathbf{p} + \mathbf{q} \neq \mathbf{o}$. Let ε be a positive number, and take

$$r = (1 + \varepsilon) f(\mathbf{p}) > 0,$$

$$s = (1 + \varepsilon) f(\mathbf{q}) > 0.$$

Then $f(r^{-1}\mathbf{p}) < 1$ and $f(s^{-1}\mathbf{q}) < 1$; therefore, $r^{-1}\mathbf{p}$ and $s^{-1}\mathbf{q} \in \mathcal{K}$, and by convexity

$$\left(\frac{r}{r+s}\right)r^{-1}\mathbf{p} + \left(1 - \frac{r}{r+s}\right)s^{-1}\mathbf{q} = \frac{\mathbf{p}+\mathbf{q}}{r+s} \in \mathscr{K}.$$

Thus,

$$f\left(\frac{\mathbf{p}+\mathbf{q}}{r+s}\right) \leq 1,$$

or

$$f(\mathbf{p}+\mathbf{q}) \leq r + s = (1 + \varepsilon)\{f(\mathbf{p}) + f(\mathbf{q})\}.$$

But this last inequality holds for every $\varepsilon > 0$, so $f(\mathbf{p}+\mathbf{q}) \leq f(\mathbf{p}) + f(\mathbf{q})$.

EXERCISES

47-1. Prove the union of any number of open sets is open, the union of a finite number of closed sets is closed. Take complements and obtain corresponding results for intersections.

47-2. Prove the intersection of any number of convex sets is convex, and the intersection of bounded sets is bounded. What can be said about the union of convex sets? of bounded sets?

47-3. If $f(\mathbf{p}) = f((p_1, p_2)) = \max (|p_1|, |p_2|)$, prove f is a csdf. Describe the set $f(\mathbf{p}) \leq 1$.

47-4. Decide whether each of the following is a csdf, and describe the set $f(\mathbf{p}) \leq 1$.
   (a) $f(\mathbf{p}) = f((p_1, p_2)) = |p_1 + p_2|$
   (b) $f((p_1, p_2)) = |p_1|$
   (c) $f((p_1, p_2)) = (p_1^3 + p_2^3)^{1/3}$
   (d) $f((p_1, p_2)) = (p_1^4 + p_2^4)^{1/4}$
   (e) $f((p_1, p_2)) = (p_1^2/a^2 + p_2^2/b^2)^{1/2}$, a, b constants

47-5. If $\mathscr{S}$ is the parallelogram with vertices $3\mathbf{a} + 2\mathbf{b}$, $-2\mathbf{a} + 4\mathbf{b}$, $-3\mathbf{a} - 2\mathbf{b}$, and $2\mathbf{a} - 4\mathbf{b}$, what is the associated csdf?

47-6. Suppose $\mathscr{S}$ is a bounded region and Λ is any integral lattice. Prove $\mathscr{S}$ contains at most a finite number of points of Λ.

48. The Theorems of Minkowski

Definition. *If Λ is an integral lattice and $\mathscr{S}$ is a set in $\mathbf{E}_2$, then Λ is said to be $\mathscr{S}$-admissible if no point of Λ except possibly $\mathbf{o}$ is an interior point of $\mathscr{S}$.*

For example, if $\Lambda = \Lambda(\mathbf{a},\mathbf{b})$ and $\mathscr{S}$ is the circular disc with center at $\mathbf{o}$ and radius $\frac{1}{2}$, then Λ is $\mathscr{S}$-admissible; if $\mathscr{S}$ has center at $\mathbf{o}$ and radius 1, Λ is still $\mathscr{S}$-admissible; if $\mathscr{S}$ has radius 1 and center at $\frac{1}{2}\mathbf{a}$, then Λ is not $\mathscr{S}$-admissible because $\mathbf{a}$ is an interior point of $\mathscr{S}$.

If $\mathscr{S}$ is any set in $\mathbf{E}_2$ and $\mathbf{p}$ is any point in $\mathbf{E}_2$, we use the notation $\mathscr{S} \oplus \mathbf{p}$ for the set $\{\mathbf{q} + \mathbf{p} : \mathbf{q} \in \mathscr{S}\}$. If Λ is an integral lattice and t is any real number, we define $t\Lambda = \{t\mathbf{p} : \mathbf{p} \in \Lambda\}$.

Theorem 48.1. (*Minkowski.*) *Let f be a csdf, Λ an arbitrary integral lattice, and*

$$\mathscr{S} = \{\mathbf{p} \in \mathbf{E}_2 : f(\mathbf{p}) \le 1\}.$$

Then Λ is $\mathscr{S}$-admissible if and only if there are no interior points in the intersection of any two convex bodies $\mathscr{S} \oplus \mathbf{p}$, $\mathbf{p} \in 2\Lambda$.

Proof. Notice that the bodies $\mathscr{S} \oplus \mathbf{p}$ are convex bodies since these are just the translations of $\mathscr{S}$, and the topological properties of $\mathscr{S}$ are invariant under translation. For the proof of the theorem, let us assume there is an interior point $\mathbf{q}$ in some two bodies $\mathscr{S} \oplus \mathbf{p}_1$ and $\mathscr{S} \oplus \mathbf{p}_2$, $\mathbf{p}_i \in 2\Lambda$, $\mathbf{p}_1 \ne \mathbf{p}_2$. Since $\mathbf{q} = \mathbf{c}_i + \mathbf{p}_i$, $\mathbf{c}_i \in \mathscr{S}$, we see that $\mathbf{q} - \mathbf{p}_1$ and $\mathbf{q} - \mathbf{p}_2$ are interior points of $\mathscr{S}$. Then by Theorem 47.1,

$$f(\mathbf{p}_1 - \mathbf{q}) < 1 \quad \text{and} \quad f(\mathbf{p}_2 - \mathbf{q}) < 1.$$

Since $\mathbf{p}_i \in 2\Lambda$, the point $\mathbf{p} = \frac{1}{2}\mathbf{p}_1 - \frac{1}{2}\mathbf{p}_2 \in \Lambda$, and $\mathbf{p} \ne \mathbf{o}$. But

$$f(\mathbf{p}) = \frac{1}{2}f(\mathbf{p}_1 - \mathbf{p}_2) = \frac{1}{2}f((\mathbf{p}_1 - \mathbf{q}) - (\mathbf{p}_2 - \mathbf{q}))$$
$$\le \frac{1}{2}f((\mathbf{p}_1 - \mathbf{q}) + f(\mathbf{p}_2 - \mathbf{q})) < 1,$$

so $\mathbf{p}$ is an interior point of $\mathscr{S}$. Therefore, Λ is not $\mathscr{S}$-admissible.

Conversely, suppose Λ is not $\mathscr{S}$-admissible. Then there is a point $\mathbf{q} \ne \mathbf{o}$, $\mathbf{q} \in \Lambda$, and $\mathbf{q}$ is an interior point of $\mathscr{S}$. Since both $\mathscr{S}$ and Λ are symmetric with respect to $\mathbf{o}$, the point $-\mathbf{q} \in \mathscr{S} \cap \Lambda$ and $-\mathbf{q}$ is an interior point of $\mathscr{S}$. Therefore, $2\mathbf{q} \in 2\Lambda$, $-\mathbf{q} \in \mathscr{S}$, so that $\mathbf{q} = -\mathbf{q} + 2\mathbf{q}$ is an interior point of $\mathscr{S} \oplus \mathbf{q}$, and of $\mathscr{S} \oplus \mathbf{o}$.

In preparation for proving our next theorem we must be able to find the area of the set $\mathscr{S} = \{\mathbf{p} : f(\mathbf{p}) \le 1\}$ defined by a csdf f. We find it convenient to introduce certain parallelograms which can be used in finding the area of $\mathscr{S}$. Suppose $\Lambda = \Lambda(\mathbf{p}_1, \mathbf{p}_2)$ is an integral lattice and $\mathbf{p} \in \Lambda$, say $\mathbf{p} = t_1\mathbf{p}_1 + t_2\mathbf{p}_2$. Define the parallelogram $\Pi(\mathbf{p})$ by

$$\Pi(\mathbf{p}) = \{\mathbf{q} : \mathbf{q} = u_1\mathbf{p}_1 + u_2\mathbf{p}_2, t_i \le u_i < t_i + 1 \quad (i = 1,2)\}.$$

Evidently the area of $\Pi(\mathbf{p})$ is independent of $\mathbf{p}$ and is completely determined by $\mathbf{p}_1$ and $\mathbf{p}_2$. Hence for all $\mathbf{p} \in \Lambda$, $\Pi(\mathbf{p}) = \Pi(\mathbf{o}) = \Delta(\Lambda)$.

Lemma 48.1. *Suppose $\Lambda(\mathbf{p}_1, \mathbf{p}_2)$ is an integral lattice. Suppose a real-valued function g is defined on $\mathbf{E}_2$ as follows: if $\mathbf{p} = u_1\mathbf{p}_1 + u_2\mathbf{p}_2$, then*

$$g(\mathbf{p}) = \max (|u_1|, |u_2|).$$

Then g is a csdf and for every positive k, the set

$$\mathcal{T} = \{\mathbf{p} \in \mathbf{E}_2 : g(\mathbf{p}) \le k\}$$

is a parallelogram with area $4k^2\Delta(\Lambda)$.

Proof. When we have shown that $\mathcal{T}$ is a parallelogram, it will follow by Theorem 47.2 that g is a csdf. Now, $\mathbf{p} \in \mathcal{T}$ if and only if

$$\mathbf{p} = u_1\mathbf{p}_1 + u_2\mathbf{p}_2, \quad -k \le u_i \le k \quad (i = 1,2).$$

Consider the four subsets of $\mathcal{T}$ which are the parallelograms $\Pi(\mathbf{o})$, $\Pi(-k\mathbf{p}_2)$, $\Pi(-k\mathbf{p}_1 - k\mathbf{p}_2)$, and $\Pi(-k\mathbf{p}_1 + k\mathbf{p}_2)$ of the lattice $\Lambda(k\mathbf{p}_1, k\mathbf{p}_2)$. These four subsets of $\mathcal{T}$ do not overlap; moreover, their union is $\mathcal{T}$ except for certain boundary points, so $\mathcal{T}$ is a parallelogram. Since the inclusion or exclusion of boundary points does not affect area, the area of $\mathcal{T}$ is 4 times the area of $\Pi(\mathbf{o})$ of $\Lambda(k\mathbf{p}_1, k\mathbf{p}_2)$. Therefore,

$$\text{area of } \mathcal{T} = 4|\det(k\mathbf{p}_1, k\mathbf{p}_2)| = 4k^2\Delta(\Lambda(\mathbf{p}_1, \mathbf{p}_2)).$$

Theorem 48.2. (*Minkowski.*) *Let f be a csdf, let* $\mathcal{S} = \{\mathbf{p} : f(\mathbf{p}) \le 1\}$, *and let* $A(\mathcal{S})$ *denote the area of* $\mathcal{S}$. *If* $\Lambda(\mathbf{p}_1, \mathbf{p}_2)$ *is an* $\mathcal{S}$-admissible integral lattice, then $A(\mathcal{S}) \le 4\Delta(\Lambda)$.

Proof. Let $g(\mathbf{p}) = \max(|u_1|, |u_2|)$, where $\mathbf{p} = u_1\mathbf{p}_1 + u_2\mathbf{p}_2 \in \mathbf{E}_2$. Since $\mathcal{S}$ is bounded, g restricted to $\mathcal{S}$ is bounded, so there exists a real c such that

$$g(\mathbf{p}) \le c, \quad \text{if } \mathbf{p} \in \mathcal{S}.$$

Suppose $n \in \mathcal{Z}^+$ and define

$$\mathcal{S}_n = \bigcup_{\substack{\mathbf{p} \in \Lambda \\ g(\mathbf{p}) \le n}} \{\mathcal{S} \oplus 2\mathbf{p}\}.$$

It should be clear that only a finite number of points of Λ satisfy the restriction $g(\mathbf{p}) \le n$, so $\mathcal{S}_n$ is the union of finitely many convex bodies $\mathcal{S} \oplus 2\mathbf{p}$. In fact, there are exactly $(2n + 1)^2$ points $\mathbf{p} \in \Lambda$ such that $g(\mathbf{p}) \le n$, because $g(\mathbf{p}) \le n$ if and only if $\mathbf{p} = t_1\mathbf{p}_1 + t_2\mathbf{p}_2$ with $-n \le t_i \le n$; hence there are $2n + 1$ choices for t_1 and $2n + 1$ choices for t_2.

Since Λ is $\mathcal{S}$-admissible, no two of the sets $\mathcal{S} \oplus 2\mathbf{p}$ have common interior points, so the area $A(\mathcal{S}_n)$ is the sum of the areas $A(\mathcal{S} \oplus 2\mathbf{p})$. Since $A(\mathcal{S} \oplus 2\mathbf{p}) = A(\mathcal{S})$, we have

$$A(\mathcal{S}_n) = (2n + 1)^2 A(\mathcal{S}).$$

Now $\mathbf{q} \in \mathcal{S} \oplus 2\mathbf{p}$ provided $\mathbf{q} = \mathbf{q}_1 + 2\mathbf{p}$, $\mathbf{q}_1 \in \mathcal{S}$ and $\mathbf{p} \in \Lambda$, so for all $\mathbf{q} \in \mathcal{S}_n$ we have

$$g(\mathbf{q}) = g(\mathbf{q}_1 + 2\mathbf{p}) \le g(\mathbf{q}_1) + 2g(\mathbf{p}) \le c + 2n.$$

Then $\mathscr{S}_n$ is a subset of the parallelogram $g(\mathbf{p}) \le c + 2n$ with area $4(c + 2n)^2 \Delta(\Lambda)$, so $A(\mathscr{S}_n)$ cannot exceed this area. Hence

$$(2n + 1)^2 A(\mathscr{S}) = A(\mathscr{S}_n) \le 4(c + 2n)^2 \Delta(\Lambda).$$

Divide through this last inequality by n^2, then take limits as $n \to \infty$ to complete the proof.

An easy deduction from the theorem is given in Corollary 48.2a.

Corollary 48.2a. *Let f be a csdf, $\mathscr{S} = \{\mathbf{p} : f(\mathbf{p}) \le 1\}$, and Λ any integral lattice. If $A(\mathscr{S}) > 4 \Delta(\Lambda)$, then Λ contains an interior point of $\mathscr{S}$ different from $\mathbf{o}$.*

We can also conclude Λ contains a point $\mathbf{p}$ of $\mathscr{S}$ different from $\mathbf{o}$ if $A(\mathscr{S}) = 4 \Delta(\Lambda)$, though $\mathbf{p}$ may not be an interior point of $\mathscr{S}$. The argument depends upon the fact that there is only a finite number of points $\mathbf{p} \in \Lambda$ such that $f(\mathbf{p}) \le k$ for every constant k, which is a geometrically obvious consequence of the boundedness of the body $f(\mathbf{p}) \le k$. If $A(\mathscr{S}) = 4 \Delta(\Lambda)$, then for every positive ε, consider the convex body $(1 + \varepsilon)\mathscr{S}$. Since $A((1 + \varepsilon)\mathscr{S}) > 4 \Delta(\Lambda)$, by Corollary 48.2a, Λ contains an interior point $\mathbf{p} \ne \mathbf{o}$ of $(1 + \varepsilon)\mathscr{S}$ for every $\varepsilon > 0$. But the points in $2\mathscr{S}$ all satisfy $f(\mathbf{p}) \le 2$ and there are only finitely many such $\mathbf{p} \in \Lambda$; say these points are $\mathbf{q}_1, \ldots, \mathbf{q}_n$. We will show that one of the $\mathbf{q}_j \in \mathscr{S}$.

Suppose instead that $\mathbf{q}_j \notin \mathscr{S}$, $j = 1, \ldots, n$. Then $1 < f(\mathbf{q}_j) \le 2$, so if we choose

$$\delta = \min_{1 \le j \le n} (f(\mathbf{q}_j) - 1)$$

then $\delta > 0$. There is an interior point $\mathbf{q}_i$ in $(1 + \delta/2)\mathscr{S}$, so $\mathbf{q}_i$ is of the form $t\mathbf{p}$, $\mathbf{p} \in \mathscr{S}$ and $|t| < 1 + \delta/2$. Therefore, $f(\mathbf{q}_i) = |t| f(\mathbf{p}) \le 1 + \delta/2$; but $\delta \le f(\mathbf{q}_i) - 1$, so $\delta + 1 \le f(\mathbf{q}_i) \le 1 + \delta/2$, which is a contradiction. We have proved Corollary 48.2b.

Corollary 48.2b. *Let f be a csdf, $\mathscr{S} = \{\mathbf{p} : f(\mathbf{p}) \le 1\}$, and Λ any integral lattice. If $A(\mathscr{S}) = 4 \Delta(\Lambda)$, then Λ contains a point of $\mathscr{S}$ different from $\mathbf{o}$.*

Example. Consider this problem: Let $f(\mathbf{p}) = f((p_1, p_2)) = |p_1| + |p_2|$, and $\mathscr{S}$ be the convex body determined by $f(\mathbf{p}) \le 1$. Find a constant c such that every integral lattice satisfying $\Delta(\Lambda) \le c$ contains a point $\mathbf{p} \in \mathscr{S}$, $\mathbf{p} \ne \mathbf{o}$.

The body $\mathscr{S}$ is the square with vertices $\mathbf{a}, \mathbf{b}, -\mathbf{a}, -\mathbf{b}$. Since $A(\mathscr{S}) = 2$, we want c such that $2 \ge 4c$; take $c = \frac{1}{2}$. For $\Lambda((\frac{1}{2},\frac{1}{2}),(2,1))$, $\Delta(\Lambda) = \frac{1}{2}$ and the point $(\frac{1}{2},\frac{1}{2}) \in \mathscr{S}$. Notice that $(\frac{1}{2},\frac{1}{2})$ is a boundary point of $\mathscr{S}$.

For $\Lambda((1,1),(1,\frac{1}{2}))$, again $\Delta(\Lambda) = \frac{1}{2}$, and the point $(0,\frac{1}{2}) \in \Lambda$ and is an interior point of $\mathscr{S}$.

EXERCISES

48-1. Suppose $f(\mathbf{p}) = |\mathbf{p}|$ and $\mathscr{S} = \{\mathbf{p} : f(\mathbf{p}) \le 1\}$. Decide whether the following lattices are $\mathscr{S}$-admissible:
　(a)　$\Lambda(\mathbf{a},\mathbf{b})$;
　(b)　$\Lambda(\mathbf{a}, (\pi/4, 1))$;
　(c)　$\Lambda((1/\sqrt{2}, 1/\sqrt{2}), (3, 1))$.

48-2. With $f(\mathbf{p}) = f((p_1, p_2)) = |p_1| + |p_2|$ and $\mathscr{S}$ determined by $f(\mathbf{p}) \le 1$, decide whether each of the following is $\mathscr{S}$-admissible:
　(a)　$\Lambda(\mathbf{a},\mathbf{b})$;
　(b)　$\Lambda(\tfrac{1}{2}\mathbf{a},\mathbf{b})$;
　(c)　$\Lambda((1,2),(3,2))$.

48-3. If $f(\mathbf{p}) = f((p_1, p_2)) = \max(|p_1|, |p_2|)$, and $\mathscr{S}$ is the set determined by $f(\mathbf{p}) \le 1$, find a point of $\mathscr{S}$ in each of the following lattices:
　(a)　$\Lambda(\tfrac{1}{2}\mathbf{a}, 2\mathbf{b})$;
　(b)　$\Lambda((1,2),(1,3))$;
　(c)　$\Lambda((\tfrac{1}{2},3),(\tfrac{1}{4},3))$.

48-4. Suppose α, β, γ are real, $\mathbf{p} = (x, y)$, and $F(\mathbf{p}) = \alpha x^2 + 2\beta xy + \gamma y^2$. If $F(\mathbf{p}) > 0$ for all $\mathbf{p} \ne \mathbf{o}$, F is called a *positive definite binary quadratic form*. The number $\delta = \alpha\gamma - \beta^2$ is the *discriminant* of the form. Prove F is positive definite if and only if $\alpha > 0$ and $\delta > 0$. [*Hint*: Show that there are points $\mathbf{p}, \mathbf{q}$ such that $F(\mathbf{p}) = \alpha$, $F(\mathbf{q}) = \delta$; show also that $F(\mathbf{p}) = u^2 + v^2$ for some u, v.]

48-5. If $F(\mathbf{p})$ is a positive definite quadratic form (see Exercise 48-4), and $f(\mathbf{p}) = \sqrt{F(\mathbf{p})}$, prove f is a csdf.

48-6. Suppose F, the form of Exercise 48-4, is a positive definite and f is the function defined in Exercise 48-5. Let r^2 be the least value of $F(\mathbf{p})$ for $\mathbf{p} \ne \mathbf{o}$, $\mathbf{p} \in \Lambda(\mathbf{a},\mathbf{b})$. If $\mathscr{K} = \{\mathbf{p} : f(\mathbf{p}) \le r\}$, $\mathbf{m} = (\sqrt{\alpha}, 0)$ and $\mathbf{n} = (\beta/\sqrt{a}, \sqrt{\delta}/\sqrt{\alpha})$, prove that $\Lambda(\mathbf{m},\mathbf{n})$ is $\mathscr{K}$-admissible. Hence use Theorem 48.2 to conclude that the minimum value r^2 satisfies $r^2 \le 4\sqrt{\delta/\pi}$.

49. Applications to Farey Sequences and Continued Fractions

The fractions a/b of the Farey sequence of order n can be put in a one-to-one correspondence with certain lattice points in the closed triangle $\mathscr{T}$ bounded by the lines $y = x$, $y = 0$, and $x = n$. The point (b, a) in $\mathscr{T}$ is the correspondent of the fraction a/b of $\mathscr{F}_n$ if and only if the line segment from $\mathbf{o}$ to (b, a) contains no point of $\Lambda(\mathbf{a},\mathbf{b})$ except $\mathbf{o}$ and (b, a). The order of occurrence of the fractions a/b in $\mathscr{F}_n$ is the same as the order in which the points (b, a) will be intersected by a line moving counter clockwise starting from the x-axis.

Suppose $\mathbf{p} = (b,a)$ and $\mathbf{q} = (d,c)$, and $a/b < c/d$ are successive in $\mathscr{F}_n$. Then there are no lattice points in the triangle $\mathbf{o}\,\mathbf{p}\,\mathbf{q}$ except the vertices; hence by Exercise 46-6, $\mathbf{p},\mathbf{q} \simeq \mathbf{a},\mathbf{b}$. It follows that $|\det(\mathbf{a}\,\mathbf{b})| = |\det(\mathbf{p}\,\mathbf{q})| = 1$, which is a restatement of Corollary 33.1a.

In the infinite wedge bounded by $\mathbf{op}$ and $\mathbf{oq}$, let $\mathbf{m} = (v,u)$ be the interior point of minimal length with relatively prime coordinates. Clearly, $\mathbf{m}$ is the point corresponding to the mediant of a/b and c/d. Now neither of the triangles $\mathbf{opm}$, $\mathbf{omq}$ contains interior points from Λ, so $\Lambda(\mathbf{p},\mathbf{m}) = \Lambda(\mathbf{m},\mathbf{q})$ and $|\det(\mathbf{p}\,\mathbf{m})| = |\det(\mathbf{m}\,\mathbf{q})|$; this is another proof of Corollary 33.1b.

We now turn our attention to infinite continued fractions. If ρ is irrational and $C_n = P_n/Q_n$ are the convergents to ρ, define a sequence of vectors $\mathfrak{c}_n = (Q_n, P_n)$. The geometric interpretation of $C_0 < C_2 < \cdots < \rho < \cdots < C_3 < C_1$ is that the points $\mathfrak{c}_{2k}$ are all below the line $y = \rho x$ and the points $\mathfrak{c}_{2k+1}$ are all above the line. Since $|P_n| < |P_{n+1}|$ and $|Q_n| < |Q_{n+1}|$, the vectors $\mathfrak{c}_n$ increase in length as n increases, and get arbitrarily close to the line $y = \rho x$ because $\lim C_n = \rho$. In Figure 9 the situation is illustrated for $\rho = (3 + \sqrt{5})/2 = \|2,1,1,1,\ldots\|$.

If there were an interior point $\mathbf{p} = (b,a)$ in the triangle $\mathbf{o}\,\mathfrak{c}_n\,\mathfrak{c}_{n+1}$, then a/b would be a better approximation to ρ than the convergents, and this would contradict Lemma 35.4. Hence, for every $n \in \mathscr{Z}^+$, $\Lambda(\mathfrak{c}_n, \mathfrak{c}_{n+1}) = \Lambda(\mathbf{a},\mathbf{b})$, and the area of every triangle $\mathbf{o}\,\mathfrak{c}_n\,\mathfrak{c}_{n+1}$ is $1/2$. For $n \in \mathscr{Z}^+$, we have $\mathfrak{c}_n, \mathfrak{c}_{n+1} \simeq \mathfrak{c}_n, \mathfrak{c}_{n-1}$, so from Exercise 46-3, there is an integer a such that $\mathfrak{c}_{n+1} \pm \mathfrak{c}_{n-1} = a\,\mathfrak{c}_n$. By considering lengths of these vectors, we can see we must take the minus sign if $a > 0$, and this leads to the recursive relations on the sequences $\{P_n\}$ and $\{Q_n\}$. The above geometric observations were made by Felix Klein.

We now prove a continued fraction is periodic if and only if it represents a quadratic irrational. First, assume ρ has a periodic expansion, say

$$\rho = \|a_0, a_1, \ldots, a_n, \overline{b_1, \ldots, b_m}\|,$$

where the vinculum indicates the repeating part, and we will prove ρ is a quadratic irrational. Of course, this is the trivial half of the theorem and a proof is easily obtained by following the pattern used in Exercise 32-3; if $\sigma = \|\overline{b_1, \ldots, b_m}\|$ is the purely periodic part of the expansion, then

$$\sigma = \|b_1, \ldots, b_m, \sigma\| = \frac{\sigma P_m + P_{m-1}}{\sigma Q_m + Q_{m-1}}$$

leads to a quadratic equation in σ with integer coefficients, so σ is a quadratic irrational; then $\rho = \|a_0, \ldots, a_n, \sigma\|$ gives

$$\rho = \frac{\sigma P_n + P_{n-1}}{\sigma Q_n + Q_{n-1}}$$

and by rationalizing the denominator one completes the proof.

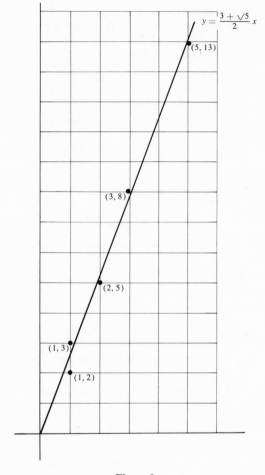

Figure 9

We prefer here to give a geometric proof that the purely periodic part σ is a quadratic irrational. In the sequence of convergents to σ, consider the subsequence $\{C_k'\}$ defined by

$$C_1' = \|b_1, \ldots, b_m\|$$
$$C_k' = \|b_1, \ldots, b_m, C_{k-1}'\|, \qquad k > 1.$$

The terms of this subsequence satisfy the relation

$$C'_{k+1} = \frac{P_m C'_k + P_{m-1}}{Q_m C'_k + Q_{m-1}}, \qquad k = 1, 2, \ldots.$$

We assume $P_m Q_{m-1} - P_{m-1} Q_m = 1$; the details of the argument are not much different if this is -1. Also, for notational ease, let $P_m = \alpha$, $P_{m-1} = \beta$, $Q_m = \gamma$, $Q_{m-1} = \delta$, and define the transformation T for all positive real x by

$$T(x) = \frac{\alpha x + \beta}{\gamma x + \delta}, \qquad \alpha\delta - \beta\gamma = 1, \qquad \alpha,\beta,\gamma,\delta \in \mathcal{Z}^+.$$

If we further define the iterates of T to be $T^2(x) = T(T(x))$, $T^k(x) = T(T^{k-1}(x))$, then we have

$$T^k(C'_1) = C'_{k+1}.$$

It is easily shown that if $x_1 < x_2$, then $T(x_1) < T(x_2)$. Also, $T(0) = \beta/\delta$ and $\lim T(x) = \alpha/\gamma$ as $x \to \infty$; therefore $\beta/\delta \le T(x) < \alpha/\gamma$ for all real x. In particular, regardless of the value C'_1, the number C'_2 is on the interval $[\beta/\delta, \alpha/\gamma)$. Furthermore, there is a *fixed point* [a number x_0 such that $x_0 = T(x_0)$] under this transformation,

$$x_0 = \frac{\alpha - \delta + \sqrt{(\delta - \alpha)^2 + 4\beta\gamma}}{2\gamma} = \frac{\alpha - \delta + \sqrt{(\alpha + \delta)^2 - 4}}{2\gamma}.$$

Notice that x_0 is real because $\beta,\gamma \in \mathcal{Z}^+$, and irrational because the discriminant $(\alpha + \delta)^2 - 4$ can be a square only if $\alpha = \delta = 1$, which is not possible $(\alpha = P_m > 1)$. From the equation $x_0 = T(x_0)$ it is clear that x_0 is a quadratic irrational; we will show $\sigma = x_0$ (see Figure 10).

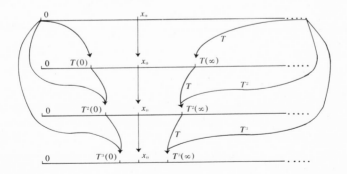

Figure 10

Consider now the difference $|T(x) - x_0|$ for any x. We have

$$|T(x) - x_0| = |T(x) - T(x_0)|$$

$$= \left| \frac{x - x_0}{(\gamma x + \delta)(\gamma x_0 + \delta)} \right| < \frac{|x - x_0|}{\gamma x_0 + \delta} < \frac{|x - x_0|}{2}.$$

In the inequality $|T(x) - x_0| < |x - x_0|/2$, we replace x by $T(x)$ and continue inductively to obtain

$$|T^k(x) - x_0| < \frac{|x - x_0|}{2^k}, \qquad k = 1, 2, \ldots.$$

With $x = C_1'$ and k sufficiently large, this proves $\{C_k'\} \to x_0$; but the sequence of all convergents approaches σ, and every subsequence must converge to the same limit. Therefore, $\sigma = x_0$ is a quadratic irrational and the conclusion is extended to ρ, as indicated previously.

Suppose conversely that ρ satisfies the equation

(8) $$ax^2 + bx + c = 0, \qquad a, b, c \in \mathcal{Z}$$

so that

$$\rho = \frac{-b \pm \sqrt{D}}{2a}$$

with $D = b^2 - 4ac > 0$, and D is not the square of an integer. We showed in Section 32 that if P_n/Q_n is the nth convergent to ρ, then the nth complete quotient θ_n^{-1} satisfies

$$\rho = \frac{P_n \theta_n^{-1} + P_{n-1}}{Q_n \theta_n^{-1} + Q_{n-1}}.$$

By solving this equation for θ_n^{-1} and using the values of ρ we find

$$\theta_n^{-1} = \frac{\rho(-Q_{n-1}) + P_{n-1}}{\rho Q_n - P_n} = \frac{\mp Q_{n-1}\sqrt{D} + 2aP_{n-1} + bQ_{n-1}}{\pm Q_n \sqrt{D} - (2aP_n + bQ_n)}.$$

We rationalize the denominator and make some algebraic simplifications to find that

(9) $$\theta_n^{-1} = \pm \frac{(-1)^{n+1}\sqrt{D} - R_n}{2S_n}$$

where

$$R_n = 2aP_{n-1}P_n + b(P_{n-1}Q_n + P_nQ_{n-1}) + 2cQ_{n-1}Q_n$$

$$S_n = aP_n^2 + bP_nQ_n + cQ_n^2.$$

Now we set up a correspondence between the numbers θ_n^{-1} in (9) and points $\mathfrak{p}_n \in \mathbf{E}_2$ by

$$\theta_n^{-1} \leftrightarrow \mathfrak{p}_n = ((-1)^{n+1}\sqrt{D} - R_n, 2S_n).$$

Although it appears that the R_n and S_n increase without bound in absolute value as n increases, we will show that the collection of points $\mathbf{p}_n$ is a finite set.

We let $\delta_n = P_n - \rho Q_n$; then

$$|\delta_n| = Q_n \left| \frac{P_n}{Q_n} - \rho \right| < \frac{1}{Q_n}$$

by Corollary 32.1; therefore we have $P_n = \rho Q_n + \delta_n$, $|\delta_n| < 1/Q_n$. Therefore,

$$S_n = a(\rho Q_n + \delta_n)^2 + b(\rho Q_n + \delta_n)Q_n + cQ_n^2$$

$$= (2a\rho + b)\delta_n Q_n + a\delta_n^2$$

where we have used (8) to eliminate the terms containing Q_n^2. Taking absolute values, we find

$$|S_n| \le |2a\rho + b| + \frac{|a|}{Q_n^2} \le |2a\rho + b| + |a|;$$

notice that this bound is independent of n. Next we observe that $R_n^2 - 4S_{n-1}S_n = D$ for all n, and since S_n and S_{n-1} have the same upper bound we see that

$$|R_n^2| = |D + 4S_{n-1}S_n| \le D + 4(|2a\rho + b| + |a|)^2,$$

and again the bound is independent of n.

Suppose $\mathscr{A}$ is the set of all points $\mathbf{p}_n$; let $\lambda \subset \Lambda(\mathbf{a},\mathbf{b})$ be the set $\lambda = \{-R_n\mathbf{a} + 2S_n\mathbf{b}\}$. Then

$$\mathscr{A} \subset \{\lambda \oplus (\sqrt{D},0)\} \cup \{\lambda \oplus (-\sqrt{D},0)\}.$$

Since there are only finitely many choices for the R_n and S_n, λ is a finite set and so is the set of translations $\lambda \oplus (\pm\sqrt{D},0)$; hence $\mathscr{A}$ contains only a finite number of points $\mathbf{p}_n$ so there are only finitely many distinct complete quotients θ_n^{-1}. It is then clear from the continued fraction algorithm that the expansion is periodic. We have proved Theorem 49.1.

Theorem 49.1. *A real number is a quadratic irrational if and only if its continued fraction expansion is infinite and periodic.*

EXERCISES

49-1. Supply the details for the proofs of Corollary 33.1a and Corollary 33.1b following the indications given in this section.

49-2. If $\mathbf{c}_n$ is the point corresponding to the nth convergent C_n in the expansion of ρ, we obtained the relation $\mathbf{c}_{n+1} \pm \mathbf{c}_{n-1} = a\mathbf{c}_n$, $a \in \mathscr{L}$. Prove that the $+$ sign cannot be used, and that $a = a_n$, the nth partial quotient of ρ.

49-3. Suppose $\alpha, \beta, \gamma, \delta \in \mathscr{L}^+$, $\alpha\delta - \beta\gamma = -1$, and

$$T(x) = \frac{\alpha x + \beta}{\gamma x + \delta}, \qquad x > 0.$$

Is this transformation monotonic? Is there a fixed point under T? If x_1 is arbitrary but fixed, does the sequence $\{T^k(x_1)\}$ converge?

BIBLIOGRAPHY

1. Harold Davenport, *The Higher Arithmetic.* Harper, New York, 1960.
2. Emil Grosswald, *Topics from the Theory of Numbers.* Macmillan, New York, 1966.
3. G. H. Hardy and E. M. Wright, *An Introduction to the Theory of Numbers,* 4th ed. Oxford University Press, London, 1959.
4. William J. LeVeque, *Topics in Number Theory,* Vol. I. Addison-Wesley, Reading, Mass., 1956.
5. Kurt Mahler, *Lecture Notes on Geometric Number Theory.* Notre Dame University, 1962.
6. Trygve Nagell, *Introduction to Number Theory,* 2nd ed. Chelsea, New York, 1964.
7. Ivan Niven and Herbert S. Zuckerman, *An Introduction to the Theory of Numbers,* 2nd ed. Wiley, New York, 1966.
8. Hans Rademacher, *Lectures on Elementary Number Theory.* Blaisdell, Waltham, Mass., 1964.
9. Harold N. Shapiro, *Lecture Notes on the Theory of Numbers.* New York University, 1951–1952.
10. James E. Shockley, *Introduction to Number Theory.* Holt, Rinehart and Winston, New York, 1967.

INDEX